T0366552

Colonial Trauma

Critical South

The publication of this series is supported by the International Consortium of Critical Theory Programs funded by the Andrew W. Mellon Foundation.

Series editors: Natalia Brizuela and Leticia Sabsay

Leonor Arfuch, *Memory and Autobiography*
Paula Biglieri and Luciana Cadahia, *Seven Essays on Populism*
Aimé Césaire, *Resolutely Black*
Bolívar Echeverría, *Modernity and "Whiteness"*
Celso Furtado, *The Myth of Economic Development*
Eduardo Grüner, *The Haitian Revolution*
Karima Lazali, *Colonial Trauma*
María Pia López, *Not One Less*
Pablo Oyarzun, *Doing Justice*
Néstor Perlongher, *Plebeian Prose*
Nelly Richard, *Eruptions of Memory*
Silvia Rivera Cusicanqui, *Ch'ixinakax utxiwa*
Tendayi Sithole, *The Black Register*

Colonial Trauma

A Study of the Psychic and Political Consequences of Colonial Oppression in Algeria

Karima Lazali

Translated by Matthew B. Smith

polity

Originally published in French as *Le trauma colonial. Une enquête sur les effets psychiques et politiques contemporains de l'oppression coloniale en Algérie* © Éditions La Découverte, Paris, 2018

Polity Press
65 Bridge Street
Cambridge CB2 1UR, UK

Polity Press
101 Station Landing
Suite 300
Medford, MA 02155, USA

ISBN-13: 978-1-5095-4102-7 – hardback
ISBN-13: 978-1-5095-4103-4 – paperback

A catalogue record for this book is available from the British Library.

Library of Congress Cataloging-in-Publication Data

Names: Lazali, Karima, author. | Smith, Matthew B., translator.
Title: Colonial trauma : a study of the psychic and political consequences of colonial oppression in Algeria / Karima Lazali ; translated by Matthew B. Smith.
Other titles: Trauma colonial. English | Study of the psychic and political consequences of colonial oppression in Algeria
Description: English edition. | Cambridge, UK ; Medford, MA : Polity, 2020. | Series: Critical South | "Originally published in French as Le trauma colonial. Une enquête sur les effets psychiques et politiques contemporains de l'oppression coloniale en Algérie © Editions La Découverte, Paris, 2018." | Includes bibliographical references and index. | Summary: "A powerful account of the subjective dimension of colonial domination"-- Provided by publisher.
Identifiers: LCCN 2020026070 (print) | LCCN 2020026071 (ebook) | ISBN 9781509541027 (hardback) | ISBN 9781509541034 (paperback) | ISBN 9781509541041 (epub) | ISBN 9781509545780 (adobe pdf)
Subjects: LCSH: Algeria--Colonization--Psychological aspects. | France--Colonies--Africa--Psychological aspects. | Algerians--Mental health. | Psychoanalysis and colonialism. | Political violence--Algeria--History. | Post-traumatic stress disorder--Algeria.
Classification: LCC DT295 .L32513 2020 (print) | LCC DT295 (ebook) | DDC 965/.03--dc23
LC record available at https://lccn.loc.gov/2020026070
LC ebook record available at https://lccn.loc.gov/2020026071

Typeset in 10 on 12pt Sabon
by Fakenham Prepress Solutions, Fakenham, Norfolk NR21 8NL
Printed and bound in Great Britain by TJ Books Limited

For further information on Polity, visit our website:
politybooks.com

To my son, Badri.

Each generation must discover its mission, fulfill it or betray it, in relative opacity.

Frantz Fanon, *The Wretched of the Earth*

Contents

Foreword

Mariana Wikinski

> In their ways of arresting time and encompassing space, these novels are not only irreplaceable tools of contextualization; they also create meaning out of the opacity of this colonial war and its afterlives. How can historians do their work without having read them?
>
> Benjamin Stora, from his preface to the book *Memoria(s) de Argelia. La literatura francófona-argelina y francesa al servicio de la historia*

> When children hear the voice of the dead, these are most often the voices of those who died without burial, without a rite.
>
> Lionel Bailly, quoted by Françoise Davoine and Jean-Max Gaudillière in *History beyond Trauma*

Suddenly a phrase interrupts the rhythm of my reading and forces me to pause. Its familiarity surprises me: "*to kill death*." "Matar la muerte": this is the title of a text published in Buenos Aires in 1986 by the Argentine psychoanalyst Gilou García Reinoso and cited by Karima Lazali in its French translation (as "Tuer la mort," 1988). I begin this foreword in what might be an excessively self-referential way, inevitably, in order to give an account of what it meant for me to feel such a surprisingly strong sense of familiarity in the very place where I was expecting to undergo a certain estrangement.[1] My practice as an Argentine psychoanalyst, working with a human rights organization and in a post-dictatorship context since 1984, and Karima Lazali's practice, working in Paris and Algeria since 2002 and 2006, respectively, converge all of a sudden in this eloquent phrase, which alludes to the wretched phenomenon that is the systematic disappearance of persons. "*To kill death*" thus functions symbolically as a historical and geographical bridge between two experiences that are politically very different: the Argentine and the Algerian. And it is precisely the differences between these experiences that prompt two acknowledgments. The first of these is that, anywhere on the planet, the setting to work of a psychoanalytic apparatus requires us to think the subject in the context of its moment and its historical and political determinations, to prevent *blanks* in the subject's psychic constitution from being replicated in the form of "holes" within the therapeutic process.[2] The second acknowledgment is that the effects of the systematic method involving the disappearance of persons, both in Algeria and in Argentina, have been devastating; we are dealing with a biopolitical tool of domination, a tool for the control of subjectivities and bodies in systems of terror. With profound sensitivity, Lazali shows that the disappearance of persons always generates an erasure at the level of memory that cuts across generations and corrosively impedes the work of mourning.

But does this allow us to conclude that Argentina's history and Algeria's are somehow homologous? Definitely not. In this, her second book, written after *La parole oubliée* (2015), Lazali unravels the elements of Algeria's specificity: the traces of trauma and the psychic transpositions of the destruction that French colonialism left in Algerian society.

French colonialism and its dramatic historical consequences – events that were scandalous in their scale, their cruelty, and their persistence – left an indelible mark on Algerian history. This history is also marked – and this makes it radically different from the history of Argentina – by the absence of investigations into and justice for the innumerable crimes committed under colonialism, during the War of Liberation and the civil wars, and even today: disappearances, genocide, the mutilation of bodies, expropriations, and the disappearance of children. These are deaths, Lazali indicates, that are deprived of bodily integrity, becoming unrecognizable. The disappearance of persons is thus not only a matter of the spectral condition of what cannot be seen; it also results from what is excessively visible but not identifiable: disfigured and mutilated bodies, deprived of any possibility of being granted an identity.

If we could think of reality itself as a laboratory functioning at the planetary scale, then comparing the subjective effects of the systematic disappearance of persons in two countries, Argentina and Algeria, might offer us definitive proof of the place of thirdness that justice creates in the ordering of social bonds. As is well known, in Argentina, the Trial of the Juntas (*Juicio a las Juntas Militares*) began in 1985, under the new democracy formed after the end of the dictatorship – which lasted from 1976 to 1983, used torture as a systematic method for social control, and disappeared 30,000 people. The trial led to the sentencing of commanding officers. The trials of hundreds of others responsible for state terrorism during the dictatorship continue even today (marked by interruptions and political vicissitudes that are too numerous to discuss in detail in this context), in cases of crimes against humanity that are still pending or that have already concluded in cities throughout the country. This process prompted me to write the following about the statements made during these trials:

> "And one day they didn't come back," some witnesses say, family members of the disappeared. … But when, *on what day*, did they not come back? How can we indicate *that day*, if all days until today are in fact that day? How can we define the absence of absence? Can we understand that those responsible are being tried for "disappearances" and not for "deaths"? What do we need in order to name the unnameable, identify the unidentifiable, specify the unspecifiable, locate the unlocatable? How can we date and provide coordinates for what never took place? How could the witness acknowledge the existence of a crime that was never definitively committed, because it keeps being committed? (Wikinski, 2016, p. 88)

In Algeria, as Lazali explains, the fate of the bodies that were torn apart or disappeared has never been investigated. Nor have there been investigations to determine who was responsible during each of the phases in which these crimes took place.

Lazali lucidly describes terror as a *psychic state* that, unlike trauma, does not allow for forgetting or repression, that does not lead to the emergence of a new subjective position, but instead blurs the boundaries between the psychic apparatus and the biological body, between the singular and the collective, between the inside and the outside. Terror remains untethered; it cannot be circumscribed.

It is perhaps a matter of an encrustation without a subject, of a devastation that can even prevent the recognition of the state of terror by the subject who is undergoing and suffering from it.

The author can distinguish between trauma and terror in this way and can demonstrate that the notion of trauma is insufficient for explaining the effects of colonial violence because she fluently traverses the fields of subjective singularity, collective phenomena, the clinic, literature, and politics, and because she clearly identifies present, historical, and trans-subjective phenomena. In this way, she reveals the traces of colonialism on both the social and subjective levels, considering an event that was by all accounts unlimited in its effects, one that resounds deafeningly in the subjective journeys of many generations throughout Algerian history. We do not find in Lazali's work any fictitious distinction between the individual and the collective; nor do we find a failure to distinguish between these realms. Instead, we confront a profoundly Freudian way of thinking in which an articulation between these spaces is constantly produced, in a fabric of numerous determinations that are always interwoven with one another.

In this sense, Lazali can be seen to be indebted to the work of Frantz Fanon, who, writing while events were still unfolding, was able to address the implantation of alienation in the psyches of the colonized as well as the improbable work of subjective decolonization that it entailed.

There will always be an expropriation of the self when colonization is imposed. What cannot be spoken of in the space of psychoanalysis, the subjective dimension in history, the unthought that finds expression in literature: it is in these recesses, Lazali tells us, that we can perhaps find the keys for understanding the subjective effects of a history of devastation whose beginning will soon mark its two-hundredth anniversary.

Many figures of the negative – negation, denegation, foreclosure, the "hole," disavowal, repudiation – can be adduced to give an account of this blank, or what is at times, according to Lazali, a "black silence" that marks what "it is impossible to forget."

The author refers to Francophone Algerian literature – a corpus that includes several clearly autobiographical works – to glean what cannot be said from the *critical deviations* (*détournements*) of this literature's language and from its use of transliteration. Deviation for its own sake becomes a value, a process that makes "the untranslatable" into an object to be transmitted. Lazali also finds resources in these novels that can be used to oppose the censorship of thought and language that has marked colonial and postcolonial history. "How could psychoanalysts work without having read them?" Lazali might thus wonder, paraphrasing Benjamin Stora.

Lazali reads the works of Kateb Yacine, Nabile Farès, Jean El Mouhoub Amrouche, Malek Haddad, Yamina Mechakra, Chawki Amari, Rachid Mimouni, Mansour Kedidir, Mohammed Dib, Samir Toumi, Amin Zaoui, Kamel Daoud, Mouloud Feraoun, and Albert Camus in order to shed light on the zones rendered invisible by colonialism and by the leveling "*mise sous totalité*" of postcoloniality.

If, having read this extraordinary book, we had to choose *one* word with which to express compellingly the effects of colonialism on both sides of the Mediterranean, we would surely choose the French word *effacement*, meaning erasure. This is an erasure that is political in its origins, of course, begun by French colonialism, with its need to deny the fact that it deposited this abject remainder of the monarchy, which was "exported" to the colony. Identity, language, tradition, genealogy, patronyms: all were demolished as if Algeria had no history. But Lazali suggests that the erasure also includes internal and fratricidal confrontations and postcolonial state terror. This was an erasure or non-inscription, then, of all genealogy, alterity, and difference, for, Lazali suggests, such heterogeneity threatens the work of constructing a "we," the coerced effort to create a uniform national essence or way of being that begins with the Algerian War of Liberation.

The celebration of the figure of the hero or martyr in the War of Liberation offers nothing more than an alibi, a distraction from the intensification of this erasure. This was a matter of refounding Algeria, erasing the colonial past, not "deconstructing" but rather "reconstructing," Lazali indicates. This reconstruction presupposes the creation of a heroic gesture of liberation, and it presumes that deaths caused by internal wars should not be surveyed. The postcolonial imposition of one language, Arabic, and of one religion, Islam, in Algeria led to the production of a myth: the myth of the birth of a nation that, again, arbitrary and denialist, erases the past and seeks to establish a point of origin or degree zero for history.

Perhaps this blank in the history of a colonizing (republican?) France, which denies the shameful, monarchic remainder that determined its strategy for occupation, corresponds to a historical blank in Algeria, a denial of the shameful history of responsibility for internal wars. This leads to a paradoxical effect on the way to liberation: the historiography of Algeria at first refers almost exclusively back to colonization, and a pure, urgent, extreme, and totalizing nationalism emerges to heal the damage done. As Lazali explains, an *excess of memory* also emerges in relation to colonialism, an excess that is in the service of erasure, that safeguards the blank in memory itself, like a spotlight that sheds *too much* light and so dazzles us. Thus a gradual transformation takes place: colonial trauma becomes social trauma.

Lazali writes:

> With a political agenda predicated on eradicating all forms of alterity, coloniality has inflamed hatred by seeking to preserve the One by killing the Other. To what extent does the rise of "nationalism" in Algeria coincide with the barring of alterity? And what impact has colonialism's negation of the paternal function had on contemporary politics? (p. 101)

These turn out to be central questions for the development of her argument. The advent of colonialism destroyed the paternal function, defined as a *symbolic function* and a *function of thirdness* that organizes the social bond, genealogy, the delineation of communities of belonging, and the constitution of identity on the basis of the assignment of a name. This function has been systematically and deliberately obstructed in Algeria since the historical break represented by colonialism. Deaths, disappearances, and the changing of names have made it impossible to determine who is who, whose child is whose, whose sibling is whose.

The advent of this disaffiliation led not only to the fragmentation of the social body by France and the War of Liberation. The internal wars that marked Algeria's history during the War of Liberation and continued after it, reaching their height in the 1990s, compel us to ask what the model for such incessantly repeated killing among brothers might be, and what might account for this ongoing search for and repeated removal of the figure of the father. Lazali critically revisits Freud's theory, developed in *Totem and Taboo* (she mentions James Jasper Atkinson's competing theory as well), and she wonders why the removal of the father and founder of Algerian nationalism, Messali Hadj, and the killing of Ramdane Abane (a leader in the National Liberation Front [*Front de Libération Nationale*, or

FLN]) in 1957 did not result in the formation of an alliance among brothers, but instead resulted in a bitter and fratricidal internal war that brutally pitted the FLN against Messali Hadj's followers and FLN combatants against one another, leading to a series of killings and ousters throughout Algeria's subsequent history.

In *Stasis: Civil War as a Political Paradigm* (2015), Giorgio Agamben accounts for the development of civil wars by referring to a permanent and unresolvable tension between the *oikos* (the house, the family) and the *polis*, in which civil war functions as a threshold between politicized family relations and the *polis* redefined in familial terms. It would seem that in this case of conflict between "us" and "them," at this threshold of difference and foreignness, a particular instance of what Agamben calls the *irresolvable* is at work. This is no longer the irresolvable tension between an "inside" (*oikos*) and an "outside" (*polis*), but rather a tension between a precolonial heterogeneity, excessively open to colonial invasion, and a reactive *us* whose formation required the suppression of even the most minimal divergence from the aim of constructing an illusory, unbreakable One.

As long as heterogeneity, otherness, and the foreign appear as threats, the social and subjective effects will be incalculable. This is not only because the Other will always be defined as an enemy, will always be regarded with suspicion, but also because, Lazali suggests, otherness instills psychic functioning, and thus the Other within is also experienced as a threat and as an obstacle. Lazali cites Albert Memmi, who argues, in perfect agreement with Fanon, that the task of subjective decolonization, the eradication of both the part of the self that is colonizing and the part that is colonized, is the tragic destiny of the colonized subject. Neither the colonizing nor the colonized part of the self belongs entirely to the self. This is a matter of a doubling at the level of identity that never produces mixing or confluence but instead leads to dissociation.

How, then, Lazali wonders, could we be the inheritors of what preceded our existence and what we cannot speak of, for reasons we do not know? *Hogra* (an insult, the humiliation that resulted from colonialism and crystallized its effects) is thus necessary as a signifier that gives shape to history. But it also fulfills an aiding and abetting function in that it persists, unaltered and unmodifiable, in the psychic life of future generations. Ultimately, Lazali asks, wasn't this what colonialism sought to achieve? Wasn't this the mental territory and the language of generational transmission that colonialism sought to occupy?

> The customary tools of psychoanalysis are thwarted, since, in this regard, the subject of speech, even in the sense of repression, has not been constituted. What is at stake, then, is precisely the coming into being of the subject, the subject of a history not so much censored as erased, reduced to nothing, and yet inevitably existing. (Davoine and Gaudillière, 2004, p. 47)

Samir Toumi's novel *L'Effacement*, which Lazali cites, was published in 2016. In it, Toumi, a young writer born six years after the end of the War of Liberation, gives an account of the impossibility of appropriating and of transforming the voids and erasures that, transmitted from one generation to another, remain inscribed as pure repetition outside the "interpreting apparatus" of the receiving subject.

Analyzing postcolonialism and the role of a particular form of Islamism in the eradication of the traces of the colonial, Lazali enters a symbolic world that is enormously complex, one in which language, religion, and politics mutually determine one another, in a superimposition that Lazali condenses in the name that she gives to this apparatus: the *LRP*. In this way, she analyzes the power of the apparatus to shape the psyche: Islamism's religious morality becomes a substitute for politics in its regulation of what is permitted and what is forbidden, what is thinkable and what is unthinkable, such that the figure of the citizen blurs into the figure of the believer.

We know that language does not reflect but rather constitutes thought. As Lazali explains, the apparatus of the LRP operates at an intrapsychic level, such that it is not possible to distinguish, in analysands, between social and internal prohibitions. The analyst must approach the work of analysis mindful that the subject protects its most intimate thoughts from confiscation, in order to prevent them from appearing in free association. Religious morality and psychic censorship overlap such that it is not possible to determine whether subjects ultimately speak for themselves or are spoken by the community to which they belong.

In a lucid assessment, Lazali reveals a psychic alibi or displacement that replaces an "inner revolution" (the uprising that the subject stages against itself) with another, already completed, revolution:

the War of Liberation. This subject's only oppressor is the oppressor from whom it has already been freed.

The Argentine psychoanalyst Silvia Bleichmar[3] (2009) distinguishes between two concepts that she also defines: the *constitution of the psyche* and the *production of subjectivity*. The former refers to the universals that contribute to psychic constitution (the unconscious, repression), and the latter names the historical processes that determine the constitution of the *social* subject. These latter processes are articulated with the processes of psychic constitution as well as with social, ideological, historical, and political variables. According to this description, the apparatus of the LRP would operate at the level of the *production of subjectivity*.

Every subject enters a *narcissistic contract* (Aulagnier, 2001)[4] with the family group into which it is born, but especially with the social body that gives shelter to, and that constitutes by cathecting, its subjects. This contract will become a link in and guarantee of the generational chain as long as the subject bears a sense of filiation, belonging, or social continuity. The psychic constitution of the *infans* takes place in a socio-cultural space that transcends the space of the family and makes "foundational statements," which constitute the infrastructure of the social group that shelters him or her. These statements can be mythic, scientific, or sacred. The discourse of the sacred especially locates the origin and end of the social body in one and the same place: the place of eternal truth.

> [F]rom his coming into the world, the group cathects the infant as a future voice that will be asked to repeat the statements of a dead voice and thus guarantee the qualitative and quantitative permanence of a body that will continuously regenerate itself. (Aulagnier, 2001, p. 111; translation modified)

As I have already indicated, in Algeria's history, the guarantee of continuity at the level of filiation is broken by colonialism and its aftermath, by the disappearances, the fragmentation of bodies, the changing of names, the assaults on tribal forms of belonging, and the persistence of disaffiliating practices implemented by the wielders of political power after the War of Liberation. I wonder, then, if the apparatus of the LRP might operate *in the place of*, might function as a substitute for, the chain of filiation as the apparatus that guarantees the subject's narcissistic contract with the society to which it belongs, even while this apparatus conditions the rules governing the production of thought and the subject's psychic life.

It was Piera Aulagnier (1984) who described the state of alienation as the destination and destiny of the ego's thinking function, of the ego as it seeks to eliminate all conflict and psychic suffering, including conflict between the ego and its ideals and between the ego and its desires. A step short of psychic death, the state of alienation presupposes that the subject has decathected from thought inasmuch as thought is experienced as a risk. The narcissistic contract, the apparatus of the LRP, the state of alienation, and the state of terror might thus converge in establishing the categories of persecutor and persecuted as ways of organizing intra-psychic life and the social bond; they might converge in making suspicion decisive for the subject's relation to alterity, sustaining an effort to banish from the psyche all forms of conflict that might lead the subject to a confrontation with itself or to a confrontation with the world in which it lives. The forbidden governs both the subject's knowledge of external reality and its knowledge of psychic reality, Aulagnier suggests.

In these pages, I have tried to locate the specificity of the colonial trauma that Lazali analyzes with such clarity and sensitivity, a trauma that inescapably affects subjectivity, the social bond, and the practice of psychoanalysis in Algeria. And yet for all the specificity of Lazali's framework, throughout my reading of her extraordinary book I saw how close our experiences are to one another, as if we lived in the same social space and spoke the same language. If in all colonization we see an apparatus for suppression and the domination of difference at work, in this text, by contrast, we find an ethics of hospitality, an openness to the foreign and the other that gives us the sense of being sheltered and of offering shelter to an experience of contact with alterity. If this were to leave a lasting trace in our thought, it would undoubtedly work against the repetition of such a devastating history.

Buenos Aires, February 2020

Translated by Ramsey McGlazer

References

Agamben, Giorgio (2015). *Stasis: Civil War as Political Paradigm*. Trans. Nicholas Heron. Stanford, CA: Stanford University Press.

Aulagnier, Piera (1984). *Les destins du plaisir: aliénation, amour, passion*. Paris: Presses Universitaires de France.

Aulagnier, Piera (2001). *The Violence of Interpretation: From Pictogram to Statement*. Trans. Alan Sheridan. Hove: Taylor & Francis.

Bleichmar, Silvia (2009). *El desmantelamiento de la subjetividad. Estallido del Yo*. Buenos Aires: Topía Editorial.

Davoine, Françoise and Gaudillière, Jean-Max (2004) *History beyond Trauma*. Trans. Susan Fairfield. New York: The New Press.

García Reinoso, Gilou (1986). "Matar la muerte." *Revista Psyché* 1.

Wikinski, Mariana (2016) *El trabajo del testigo. Testimonio y experiencia traumática*. Buenos Aires: La Cebra.

Introduction: The Difficulty of Acknowledging Colonial Trauma

The idea behind this book came from comparing my experiences as a psychoanalyst in Algiers and Paris. The regular tools of this exercise in self-liberation whereby the subject discovers its own forms of alienation weren't sufficient for my patients in Algeria. They couldn't turn away from the demands made upon them by the private, social, and political spheres. The notion of "resistance" doesn't adequately describe their inability to escape censorship's hold over thinking and to live fully as distinct and singular beings. Clear therapeutic benefits were present during sessions, but, as psychoanalytic treatment always goes hand in hand with a revolution of the private sphere, no matter where that treatment takes place, in Algiers this repeatedly sought-after revolution remains an unachievable goal that is systematically and tirelessly stalled by an Other: family, politics, religion ... How to go about analyzing this private sphere deprived of its revolution? And what is this melancholy-filled grievance hiding?

Although subjectivity can never be hemmed in by any identity markers – be they political, linguistic, or historical – it nevertheless uses these to weave the invisible threads of a private self. The site of psychoanalysis must be reinvented on each occasion with each new patient, while taking into consideration the various elements "saturating" the subject. Rather than over-emphasizing cultural specificity, the question raised here concerns the politics underlying a psychoanalyst's practice. It is also worth considering how treatment may shed light on a central socio-politico-linguistic dynamic at work in the larger society.

The history of French colonization in Algeria: a *blank space* in memory and politics

My psychoanalytic practice takes place between different languages (French and Arabic) and locations (France and Algeria). This has probably sharpened my awareness of difference, and made me realize what difference reveals about the reach of politics in both places. It has also made me aware of the impact of this political reach on the formation of the subject. In Paris, the fact that a vast number of French patients who, caught in a generational confusion and stagnation, evoke at some point the signifier "Algeria" invites further reflection. These French patients, usually three generations removed from colonialism, express being weighed down by a colonial history experienced more often than not by their grandparents, who were involved in either colonization or the War of Liberation, but about which these patients know very little. It is surprising to see how they are grappling with questions of shame and responsibility due to this legacy. Expressing an acute sense of discomfort, they are caught in a history they never experienced, one that, more often than not, they inherited cloaked in silence. They are beset with a number of questions: how do you inherit a past you never bore witness to and which, for unknown reasons, you can't even speak about? Where does this leave you? Where did their parents and grandparents really stand politically in relation to "coloniality," a term that covers a long period (132 years) of domination and violence, whereas now their descendants are forbidden from thinking about it? How do you develop your own story when this parental silence is met with a political *blank space*?

One might argue that Algeria crops up repeatedly in the discourse of patients because the analyst's familiarity with the matter invites it. But these patients initially came to her for a variety of symptoms that bore no *inherent* relation to this episode in History. And at some point over the course of several conversations they express the painful impression of being held hostage and left defenseless by an inaccessible past. Following the patterns traced by the signifier "Algeria" thus leads to a blank space in memory and politics. The work of historians can hardly help these patients come to terms with the ideological blind spots they inherited, for the formation of subjectivity is beyond the reach of the historical record. On the other hand, subjectivity needs and demands acknowledgment from the political order. Otherwise, the part of History refused by the political order continues to be transmitted from generation to generation and creates psychic mechanisms that entrap the subject in existential shame.

In Algeria today, the colonial question is so pervasive that we tend to think of it as a historical template. But its official history is frozen in time, one-dimensional and therefore lacking in nuance. It is a matter for politics, probably its lone and major preoccupation. Since the devastation wrought by colonialism is widely acknowledged, it is treated as though there is no point in exploring the matter any more deeply from an interdisciplinary perspective. There is no room for dispute: the matter of colonialism is, by unanimous decision, a closed affair.

The ideological blind spots shaping the current understanding of coloniality – both inside and outside of Algeria – provided the impetus for this book. The myth-filled grievances expressed by so many patients thus hide the orchestrated effort to ensure that these blind spots persist in Algeria *and* in France. In this way, each individual's responsibility in shaping History goes unquestioned. History is seen as a set of facts and the interpretation of those facts. This straitjacketed history prohibits subjects from exploring their own layered, complex, and intricate family histories. As one cannot access the history of colonialism through official channels in either France or Algeria, one must find another way in. This entails ignoring the myths and legends attached to it and seeking what lurks behind the curtain of the historical record.

History doesn't speak for itself. It speaks through subjects who, ideally, debate with historians and politicians over its interpretation. Psychoanalysis cannot do without History, and yet depending solely on it would neglect the private interpretations tirelessly made of it by individuals. Practicing psychoanalysis in Algiers has in this respect been instrumental in understanding how subjectivity stands both within and outside History. To be clear, every subject is formed within and by History. But the subject also hides behind History in order to elude questions concerning individual responsibility. Everyone is familiar with the common refrain from patients: "My past has made me who I am." And who hasn't heard healthcare practitioners claim: "They are like that because of their past"? The subject strives continually to move beyond History and yet, instead of breaking free from it, the subject hides behind it. How, then, can one read and draw on History without letting this reading drown out the subject's own interpretation of history (both personal and collective)?

To address colonialism's impact on subjectivities in Algeria today, I have drawn on the work of historians as well as on works of Algerian literature, principally those written in French. My aim is to bring together fictional enunciations [*énonciations de fiction*] and historical statements of facts [*énoncés historiques de faits*] as nothing comes closer to the texture of subjectivity than the literary text. The turn to literature is both fruitful and necessary, since the discussions I had with patients which drove me to write this book remain protected by confidentiality – a delicate matter in Algeria, where the act of *revelation* is negatively perceived in both its religious and secular iterations. Not to mention the fact that psychoanalytic treatment remains restricted to a small minority of people who are already concerned about preserving their ability to speak openly and who remain fearful that the presumed secrecy of this encounter may be betrayed. On top of all this, colonialism finds expression through the blank spaces of thought and speech – in other words, through non-discursive acts.

When I first began my research, I was surprised to note how little clinical work there was in Algeria and France on the psychological effects of colonization, apart from that of Frantz Fanon in the 1950s, which is itself a treasure trove of information and rigorous analyses on the psycho-physical harm caused by colonization.[1] His untimely death at 36 years old in 1961, and the dearth of subsequent clinical research in this field, has resulted in an absence of studies on the lasting and wide-ranging psychological effects of colonialism. This is thus a largely unexplored and unknown territory. On the other hand, the clinical effects of colonialism haven't failed to garner attention from other fields: history, anthropology, sociology, literature. What explains this blind spot in the field of medicine and psychoanalysis? Does the impact of colonialism not merit its own analysis? Do existing notions, such as trauma, go far enough to explain colonial violence?

Probably not, if one considers that the blind spot in the psychological literature signals a larger problem: the longtime silence of leading public officials on colonial violence and its persistence today in a number of disciplines, especially in the clinical and social sciences. The logic of colonialism lives on in the thoughts, speech, and practices of former colonizers and those once deemed "*indigènes*." This logic defies treatment, in the medical sense of this term: it doesn't respond to care or examination. This holds true regardless of the subject's place within colonialism (colonizer or colonized). The matter is clearly expressed through the blank spaces found in writing from both sides of the Mediterranean. This poses a serious challenge to clinical studies, which would like to establish a set of clear issues and carefully trace them back – following visible signs – to the matter of colonial violence. No such signs exist in this case. This blank space has risen to a deafening pitch in French and Algerian society, where it can be felt by all. For the clinical psychologist, the legacy of colonialism exposes an unusual psychic phenomenon, namely, the existence of a whole field of invisible traces that, in spite of their seeming absence, give shape to subjectivities and political discourse. The clinical psychologist is forced to deal with a history deprived of archives, literally and metaphorically. It is now no longer a question of deconstruction, but one of reconstructing traces that exist outside of memory.

A much-needed interdisciplinary approach

For the time being, uncovering this history is left to historians, whose work, although indispensable, fails to account for actively troubled subjectivities. What's more, in France, once a historical record is reconstructed, it rarely gets noticed outside of its own disciplinary framework. This is in stark contrast to the multidisciplinary approach of postcolonial studies in the Anglophone world, which have remained largely inaccessible to Francophone readers, in spite of the fact that these works draw on the writing of major French-language theorists: Césaire, Sartre, Fanon, Memmi, Lacan, Derrida, Foucault. Further developed and articulated abroad, this French history is paradoxically hard to translate back home. This paradox seems symptomatic of an impossible reciprocity, which results in so much research, so many carefully crafted arguments remaining in utter obscurity.

In France, there seems to be an assumption that the history of colonization falls strictly within the purview of historians and former "*indigènes*," and therefore that they have the exclusive right to treat it. And in Algeria, that colonialism belongs entirely to those who were colonized, making the rest of us apathetic to its debates and complex history. Go along now, nothing to see here! In both places, there are unbridgeable divisions – an effect of coloniality. And this issue cannot be tackled without an interdisciplinary approach, as each discipline offers a distinct vantage point on the matter. Otherwise, by succumbing to a crude partitioning of the past, present, and future, the colonial world will continue to remain sealed and inaccessible.

In Algeria, the effects of colonialism are so embedded in the psyche of individuals that it becomes difficult to distinguish between what results from direct impact and what has formed over time into an "identity" crisis caused by its disruption of the core network of subjectivities and the social bond. Subjectivities are thus entirely suffused with coloniality. This is now accepted as an unequivocal and indisputable historical fact, which undermines the idea that the primary interpreter of History is first and foremost the subject. This is probably why the consequences of colonization appear only in public discourse as cries of pain and resentment, which target the Other of colonialism while staying mute about the impacts of History on one's personal history. In Algeria, it is as though colonization is the one and only *trauma*. Whereas in France, the notion of colonial trauma is flipped on its head and exploited by the political order: much talk is made of the "benefits of colonization" for "*indigène*" subjects. The political order thus strives to make the historical record disappear and to discount the role the subject plays in History. Here again, no clinical work has considered the specificity of these traumas and their impact on the social bond. Instead, we are stuck in the hell of this duality which allows the war to persist by other means.

Bringing together psychoanalysis, history, and literature in an attempt to discern the invisible role played by politics, this book's approach may be judged problematic from within the specialized fields of each of these disciplines. But I can think of no better way of treating the politico-subjective "matter" of coloniality in Algeria, a totality that cannot be contained by isolated disciplines. Specialists from these fields each see themselves as the best positioned to take on the wide-scale devastation of this affair. But the coalition formed in this book with these three disciplines creates a dynamic approach where each discipline informs and alters the other, the cumulative effect of which, I hope, will not be lost on any reader. This transdisciplinary configuration is also a way of mounting a defense against the blanket-statement generalities and deadly binary logic that affect anyone who has tried to take a closer look at colonialism.

Literature strives to give expression to the blank spaces and the ideological blind spots present in the historical record. Above all, it alerts the reader to how a text is continuously shaped by its invisible margins. The psychoanalyst, for her part, works to read and analyze what can be read without reading. This works only insofar as the psychologization of characters and writers can be carefully avoided. It is a matter of treating the literary text as a literal object. This is clearly at odds with most approaches to literature, in the same way that my subjective approach to history stands in contrast to the historian's objective approach. The other challenge this approach encounters is its use of psychoanalysis as a tool for understanding the political dimension of history and society. One must strive to avoid psychologizing society and/or "sociologizing" the subject. Although, as Freud says, no boundary separates the individual from the community, it remains a challenge for the psychoanalyst, whose practice is based on individual experience, to understand the community through the individual and vice versa.

Putting psychoanalysis, history, and literature to use in this way brings us the closest to what has been, and continues to be, erased by the political order, whose subjects are kept in the darkness of a sleepless and endless night. History seizes, literature writes, and psychoanalysis reads what resides in the blank space of the text's margins.

1
Psychoanalysis and Algerian Paradoxes

> I'm asking God, who, by the way, no longer believes in me, for forgiveness.
>
> Sarah Haider, 2013[1]

> One can get over the disappearance of the past.
> But we cannot recover from the disappearance of the future.
>
> Amin Maalouf, 2012[2]

Caught in a continual tension between servitude and freedom, Algerian society is a mix of contradictions. Since 1962, it has seen progress in a variety of areas: schooling for children; treatment of women, who are much more visible now in public spaces; access to free healthcare, and so on. On top of all this remains an unfulfilled desire to regain a sense of belonging after the colonial order has laid waste to one's ties to the past (destroying languages, traditions, communities). Despite this progress, pain is still constantly felt and expressed by individual subjects, regardless of their sex, language, profession, or cultural belonging. This pain, expressed by numerous patients during the course of my psychoanalytic practice in Algeria, is rarely viewed as part of a larger historical and political context. Individuals feel as though they are gasping for air, suffocating, being crushed under an unbearable weight. A sense of inertia is palpable and conveys the feeling of a "foreclosed" future.

This pain manifests itself – and is recognized – through the body. But rarely is it related to a subjective history. It is hard to ignore this physical torment when so many women and men repeatedly complain about it. It becomes bigger than the individual subject who bears it in his or her personal life and begins to speak on a public stage. The one-on-one session soon gives way to a much larger social stage where the subject feels threatened. Far from inviting participation from subjects, the social order triggers their retreat. The relation between the individual and the community is troubled. On the one hand, they are divided by a radical incompatibility, since one (the community) *almost* cancels out the other (the individual), and, on the other hand, they are bound by a profound and inextricable solidarity. This solidarity is masked by grievances and other demands such that no one can perceive the role the subject plays in the social bond that overwhelms it and hastens its disappearance. How, then, can one make what is clearly present in psyches legible and translatable when it has left no traces, apart from what remains in public memory?

Disarray of the private and public spheres

In Algeria today, individuals have long been subjected to terrible psychological, social, and political realities. The traces and consequences of this history remain to be discovered. The persistence of these traces in the present is an urgent matter, but it is drowned out by the noise *seemingly* emanating from elsewhere: an international context whose instability poses many social and economic threats, a political establishment that has been in a volatile state for many years, and, not least, a resurgence of religion that goes beyond national borders and has been behind Algeria's bloody history, a history that is always ready to re-erupt.

The analyst seeking to find traces of these catastrophes within psyches and connect them to verbal expressions of "*malvie*," or the angst of young Algerians, will be met with disappointment. She will encounter only matters of another order – economic, administrative, international – that mask the real inner despair plaguing subjects and the state. Distinguishing between the inside and outside, between private and political responsibilities, between individual and collective history, is no easy task. This gives a dizzying impression of a homogeneous, all-consuming whole. Each individual's role *as individual* in the very make-up of society is constantly effaced, while an omnipotent force that lurks in the shadows is seen as fully responsible for all subjective and social disasters.

The outer catastrophes experienced by patients are described, recorded, catalogued, but correlating them with present-day effects on subjects and the larger public has remained a struggle. It is as though a gap both held the individual and the community apart and caused them to merge. The private becomes public, and, conversely, the public is quite simply private, making them an unbroken unit. This makes it exceedingly difficult to find distinct features which could be used beneficially to mediate between them. In this all-consuming whole, catastrophes are identifiable, but their specific and exact effects on the individual and the community remain hidden. Catastrophes are constantly experienced in the present tense. Past, present, and future are *almost* indistinguishable. In other words, it is hard to make the (catastrophic) event that occurs into a significant event that can be documented in the private and political spheres.

Psychoanalysis in Algeria has slowly ventured out past its regular cultural and linguistic territory and settled into these troubled waters. The 2000s were marked by an urgency to build and repair, not by a need for deconstruction, which is frequently used in analyses of the subject. The last war (1992–2000) had just shown a seemingly unprecedented level of atrocity, robbing countless children, women, and men of their voices if not their lives. The demand for psychoanalytic treatment speaks to the need to understand and move beyond the brutality experienced during what have been called the "bloody years," the "dark years," the "red decade," the "reign of terrorism," or the "nightmare years." New questions have emerged as atrocities have spilled over into the private sphere and familiar friends can no longer be distinguished from foreign foes. External catastrophes have laid waste to inner lives, borders, languages, histories. The destruction was so vast that the conventional means of separating inside from outside proved to be no longer operational, failing at times to make any sense at all.

In this indecipherable landscape – to which I'll return shortly – appeared an element that had been buried until then and which recalled one of Freud's central insights: namely, the indissociable ties between the psyche and collective experience. Freud developed this idea as early as *Project for a Scientific Psychology* in 1885, arguing that interiority springs principally from a decisive encounter with the exterior (the environment) – this is the fundamental experience of every infant. He would later refine this idea by widening his notion of environment to include the social environment in 1913 with *Totem and Taboo*, and continued in this vein all the way up until *Moses and Monotheism* (1937), where he strove to explain how the unconscious is formed, unforgettably, before boundaries are drawn. In other words, it isn't just national borders that are artificial: it starts with the border separating interiority and exteriority for the speaking beings we all become. However, we often forget this as we continue to cling to fragile borders for reassurance. Catastrophic events can put these borders at even greater risk.

In Algeria, each individual harbors within the degeneration of the collective body whose central organ is the social order. The discourse of patients from the 2000s sheds light on how this situation directly affects the bodies of subjects, especially in light of the fact that the disaster of the war of the 1990s was compounded by natural catastrophes in the following decade: repeated earthquakes, one of which caused more than 2,000 deaths, and floods no less destructive.[3] All of this is not without consequence, as each catastrophe – although different in kind – finds itself tied to the previous one. These catastrophes are linked together by their shared belonging to the tragic sphere. And the psychological associations formed can be explained by the temporal proximity of the catastrophes and the great losses of human life occasioned by each. Tragedy of this sort marks the discourse of patients, who can be heard speaking of "an unrelenting fate," of "being condemned to catastrophe," or even of "divine punishment," which evokes the "wrath of the gods" from Greek mythology. The collision between human atrocities from the war years and the ravages wrought by nature has led to a surge in religion: prayers, women turning to the veil again and a series of other acts to "placate the gods." In both cases, between heaven and earth, God is at stake: a mysterious God called upon to shield one from natural catastrophes.

Calls for help made amid the murders and massacres during what has been deemed the "Internal War" remain unheard and unanswered. They have been drowned out by the lives lost due to natural disasters. The senselessness of human cruelty has been matched and complemented by nature's unpredictability. Failing to find explanations for these, everything appears to be ruled by chance. Questions such as "How did we get here?" and "What's behind this endless bloodshed?" – the countless dead and missing, the massacres, the savagery of it all – are like so many purloined letters.

A sense of dismay has spread and taken hold of the public at large. The line separating inside from outside, a reliable barrier in normal times, is now fragile and porous. The fabric of society is torn, plunging subjects into a quasi-permanent state of uncertainty and fear. This accounts for what I perceive to be a serious "social trauma" plaguing subjectivities, one whose causes and cures have yet to be discovered.

God's reinforcement of failing institutions

The unceasing, demonic blows of the real spare no one. Everyone is exposed to them to varying degrees. Hence the unrelenting sense of a looming danger, which is all the more troubling as the source of the trauma remains unknown: heaven or earth, inside or outside, the state or religion, and so on.

Various forms of violence are embedded and rehearsed in the social sphere. For example, the vulnerability of subjects is even more acutely felt at sites of social interaction (institutions, work, family). Their feeling of defenselessness causes them to turn inward, becoming withdrawn and disengaged in order to avoid being exposed to danger. This produces a sort of tension in a public seeking a feeling of existence: on the one hand, the social fabric is being torn to pieces from all directions and continues to grapple with the long history of its fight to become a "nation," the impacts of which are hard to measure; and, on the other hand, there is also an attempt to patch up these tears, a necessary step for moving on with one's life, but also the source of new forms of violence. The social sphere both stages and witnesses these catastrophes, but it also strives at all times to cover them up, dismissing their very existence. In so doing, it only throws matters into further disarray.

Behind this tension between tearing open and patching up is the experience of the living, which has come under attack and which deserves further scrutiny. The expression of a damaged life in the social sphere and the ability of this expression to spread and wreak havoc on a subject's future should be interrogated. For the individual subject cannot be reduced to the community. It traces its own private paths that are both within the public and unreachable at its margins. And yet serious conflicts within the larger public bar the emergence of subjectivities. An unavoidable clash arises, one that is designed to serve a political purpose. Repeat experiences of trauma have directly affected people's relation to faith, as the resurgent *visibility* of religious practices has made clear. The importance of the visible forms these practices assume cannot be overstated. The material transformation of "belief" has a very particular social function, suggesting that faith depends on its visible demonstration although it remains invisible as a private matter. Faith, once immaterial, then assumes a demonstrable material existence, one that is put on display for all to see. Does this mean that religion has become no more than an outward display? What is behind this almost physical staging of belief?

The display of divine obedience serves as a bulwark against a feeling of insecurity that views the outside world as dangerous in light of its distant and not so distant past. Freud sees the "need for religion" as deriving from the infant's experience of helplessness. This deep-seated feeling "is permanently sustained by fear of the superior power of Fate."[4] The "infant's helplessness" is a primal experience of subjectivity. Each infant experiences it before a "figure of comfort" (usually the mother or a substitute) comes to put an end to its helplessness (unpleasant feelings, hunger, cold, pain, etc.). The "figure of comfort" registers as coming from the outside. This remains stamped on the psyche. Throughout the course of a person's life, these silent traces of helplessness may be reawakened when the subject is confronted by danger. This primal experience marks an initial separation between inside and outside, and creates a welcomed and awaited experience of alterity. Indeed, the comfort and security brought to the crying infant by this figure allows it to begin to distinguish between inside and outside. This is how the subject at this early stage discovers the existence of difference. The outside becomes the source of calm for inner stirrings, discomfort, pain. The subject also learns to construct an interiority to shield off dangers emanating from the outside. It "interiorizes" the figure of a supportive Other, which henceforth will remain within it. At times, it may still call on the exterior figure when the interiority it constructed isn't enough to handle the dangers that threaten it.

This primal experience of alterity plays a decisive role as it serves as a compass for navigating between internal and external reality. The outside appears as a potential source of comfort. If this Other fails to appease the child or if he or she is malicious, it can have devastating effects on the child. The only option then is to appeal to a higher power, one that is greater than humankind, since the "trust" in humans has been effectively shattered. It is worth noting that the term "trust" has taken on a negative connotation in Algeria today: it points to a significant phenomenon that dates back to the Internal War, namely, the failure of the (internal/external) Other to be a figure of comfort and security. Regular discussions with patients confirm this widespread inability to trust others, including the institutions of healthcare, law, and education. These institutions have thus seemingly lost their status as mediators in a country whose staunchly socialist government once made healthcare, education, and legal protections accessible to all at no cost. This breakdown in trust suggests that the outside and the inside are themselves sources of constant threats, and that overcoming this is anything but straightforward.

But patching over widespread despair with an overzealous display of faith isn't enough to appease a profound feeling of danger. There is a growing sense of destitution and insecurity. Let's not forget that Algeria has a long history of turning to religion as a remedy for human despair amid a political crisis. Recent events (the Internal War, natural catastrophes, "*malvie*") have only reinforced this tendency. Mohammed Dib wrote as early as 1970: "Placing our trivial concerns in the hands of God, isn't that wonderful? Only we could come up with such insights."[5] Elsewhere, he makes his point clear: "Our desire to also put ourselves in God's place knows no bounds."[6]

An excess of religious zeal goes hand in hand with a rising sense of danger and an absence of "trust." The subject sees itself in peril with no one to turn to, and, as a result, it multiplies its offerings to a supposed divine power and demonstrates its faith in a more conspicuous manner. In this way, these visible displays of faith are like so many unanswered calls. It is as though the very lack of divine response led to a dramatic increase in the need for religion. For these demonstrations of "belief" in the social sphere rarely produce the desired results, since "God chose to let us deal with his absence and our state of abandonment on our own."[7] Is this display a way of reassuring God about His own existence while trust in Him in the private sphere is thrown into doubt?

Where can one seek help if God, who remains at the helm, struggles to steer private and public life? And where can we turn when the institutions designed to protect us run aground on the shores of despair? Would secularization make our institutions more robust and trustworthy?

The power of religion and the religion of power

Today's institutions are in a severe state of disrepair for several reasons: lack of resources, corruption, an egregiously unregulated market as national socialism gave way to an anarchic capitalism, and the need to manage the crises provoked by the Internal War and natural disasters. These institutions are perceived as the site and cause of a new danger. They undermine their own purported goals of providing aid, safety, and care. There is a profound neglect of one's basic needs, not to mention one's desires. A litany of minor, everyday complaints pervades the speech of subjects, who feel their existence has become inconsequential. In this context, many shy away from institutions out of fear and apprehension. Subjects don't see them as a source of shelter, far from it. Governed by chance and confusion, institutions have become the unregulated site of an agonizing social order.

Institutions reproduce and exacerbate the tears in the social fabric, wreaking even more untold havoc by re-enacting practices of violence. This leaves the citizen powerless. Emergency exits are blocked off, from both the inside and the outside. It is therefore up to each and every individual to use the widespread art of *gfasa*, namely, finding ways out through resourcefulness and creativity. Although at times producing astounding feats, this practice tends to further discredit institutions. With each individual potentially inventing his or her own rules and laws, this may lead to ingenious discoveries as well as various forms of abuse: corruption, authoritarianism, harassment. An already threadbare social fabric and the rapid erosion of foundational institutions signal that the function of an independent mediator is in crisis. Touchstones of tradition are vanishing, including the family, whose conventional structure has been challenged by new modes of familial organization: divorce, single-parent families, the new predominance of the nuclear family.

These violent and brutal changes occurred at a pace that left no time (in the sense of duration) for new touchstones to be established to make sense of them. The construction of the psyche in children, as developing subjects, has been affected as a result. Some child psychiatrists have observed in their private practices many parents who, finding themselves all alone, are left to struggle without any institutional support. In traditional societies, the extended family served this function, whereas in modern societies, institutions play this role. However, as previously mentioned, "trust" in these institutions has vanished in Algeria, where they are seen as instruments of oppression and in part responsible for one's internal sense of insecurity. The outside is as threatening as the inside, leaving no safe breathing space, which is precisely what institutions are supposed to offer their citizens. What's more, civil society is struggling to fill the gap left by the failure of institutions.

Thus, religion is called upon to treat several levels of despair: private, political, social. The visible display of faith supplants foundering institutions that can no longer protect and care for individuals. The "power above" is an expression rarely uttered but frequently used in one's body language, such as by raising one's eyes to the sky. This suggests two things in Algeria: God and the state. This confusion between government and divine power is nothing new. It is in fact behind the origin myth of a young Algerian nation. It is worth asking, however: what God are we talking about? Since, although the Qur'an states there is only one God, multiple Gods appear in practice and in speech, depending on the place and purpose God occupies for the speaker. Gesturing toward "a power above" suggests that the designator must be "down below." Thus, a division is created between heaven and earth, between decision-makers "from above" and those living here "below," between God and humankind. Despite this apparent configuration, the gesture indicating the "power above" more often than not is made with a knowing smile, suggesting that this is a matter of playacting. The fight over who gets to occupy the places "down below" rages on without involving the "power above." This implies that the "power above" is invisible and, notably, unidentifiable. A tension thus arises between a visibility that is at once excessive and absent, which reverses the material expressions of power and religion: on the one hand, an absent and therefore invisible political power; on the other, a religious power blindingly visible.

The decision-makers are called "them" (*houma*) and, as such, are unnameable as individuals in the decision-making process, whereas God is for His part called "Him" (*houwa*). This difficulty of naming and identifying covers everything in a thick haze. Confusion and bewilderment serve very different purposes for the individual and for those in power. The cult of the invisible is a state matter while religion is concerned with the visible. In the Algerian context, the fields of the visible and invisible are flipped in order to hide the boundaries of religion and power.

The power of religion and the religion of power merge and feed into each other. The absence of secularism cannot alone explain this reversal of the fields of the visible. To be sure, the religious signifier is exploited for matters of governance, and this contributes to the current reversal; but it is also due to a form of governance that has merged history, languages, and religion together to serve its totalizing agenda. The individual effects this has had on the public raise a serious question: how does the deliberate blurring of boundaries make use of an interstitial space – between "down below" and "the power above," between the visible and the invisible – to better remain unseen?

In this space, which belongs to no one, there is as much corruption as there is subversion. And the recent rise of an Algerian literature that explores contemporary social problems and brings the years of the Internal War under intense scrutiny illustrates the fruitfulness and dynamism of these intermediary spaces. Once again, fictional treatments of contemporary issues have paved the way for scientific analyses to be carried out in various disciplines. Novels, art, and films anticipate the cultural and theoretical work to come.

In a rather dark landscape, those leading the struggle carve out an admittedly narrow but nonetheless real path. Many stories get written between heaven and earth. In Algerian literature, writing becomes a vehicle for censored languages and confiscated dreams. It is thus a manner of giving life to an alterity that is suppressed in the political sphere. To accomplish this, Algerian writers deftly deploy the art of *détournement*.[8] In their hands, language becomes a ruse they can manipulate for their own purposes. The literary text repeatedly calls attention to the official (political) narrative in an effort to divert the censors while making their erasures visible. Writing is an antidote to political coercion. Nabile Farès (1940–2016) had noted this problem as early as 1976: "We are besieged and erased by those in power like dirty words written in sand."[9]

Contemporary Algerian literature is a deliberate refusal of the many forms of deceit first practiced by the colonial power and then by the Algerian state. In 2016, the young writer Samir Toumi illustrated in his novel *L'Effacement* the alarm of an Algerian subject who is suffering from a terrible condition: the disappearance of his reflection and the erasure of his memory. Erased and absorbed into a "triumphant" nation, this character is a metaphor for the political treatment of memory in Algeria. The suffering subject feels that "everything is being erased, inexorably, ... I've been set adrift in a hazy space where everything is blurred and *blanked out*."[10] Struggling with the disappearance of his reflection, the son calls on his father for help, but his father urges him to embrace being swept into a heroic "national" narrative:

> He told me it wasn't worth it, because I had him. He was the reflection of both my body and soul. Besides, it had always been this way. When I told him about my vanishing and the disappearance of all my memories, he shrugged his shoulders. You have mine, he responded, they are more interesting and much richer. *I have a war to give you, a magnificent victory and the building of an immense nation, what more could you want? I'm giving them to you, my memories are yours.* I thanked *papa* as he stroked my head.[11]

The literary text and the invisible staging of power

The inaugural political gesture of the first Francophone Algerian writers of the colonial era came from their *détournement* of one language (French) for the benefit of another (their mother tongue). This also gave them an original literary style, one distinguished by an invisible plurality of linguistic and textual spaces, both sanctioned and unsanctioned by the colonial order. Many languages and modes of thought were "smuggled" into these literary texts, constituting in this way their own underground space.[12] This is how *détournement* disrupts the channels between private and public censorship. This practice also exposes the staging of power. It does this by making invisible political spaces visible, thereby stripping censorship of its power by allowing it to be heard in the text. Writing gives (textual) form to what is erased or prohibited from the subject's thinking. What is silenced in an individual's speech finds expression in the literary text.

On the one hand, the novel designates the unnameable and shines light on what is held in the dark by the political order. On the other hand, on a daily basis, subjects continue to submit to the law of silence and repression, making speech the site *par excellence* of allusion and enigma, as observed in my clinical practice in Algiers. This fine-tuned operation serves many paradoxical purposes: it both helps and hurts the subject in its life decisions and thought processes. It works by construing speech, thought, and actions as a single totality. This stands in contrast to the traditional form of totalitarianism, where what is and isn't allowed is dictated by the political order and never mistaken by the subject. In this case, a totalizing force is internalized with no visible and identifiable dictator. This is why I prefer to call it *totalizing* rather than *totalitarian*.

In this context, the subject depends on artful schemes. At first, it believes itself smarter than the censors: it submits to them superficially while hoping to carve out an invisible path ("unseen, unharmed") for its dreams and desires. Submitting to the censors' laws becomes therefore a means of transgressing them. This is tantamount to saying: "Since nothing is allowed in clear daylight, then everything becomes possible for me in the dark." Except that the practice of *détournement* – which takes pleasure in disruption – proves to be short-lived and can easily turn into its opposite. Indeed, little by little, the subject begins to forget the subversive aim of its ruse. It ends up agreeing to the terms of its own imprisonment and abandons along the way its initial goal, which was to elude the censors. Brought under the totalizing force of the exterior censors it sought to subvert [*détourner*], it unwittingly helps design the system it was fighting against.

As a result, censorship becomes so ingrained that it feels like law. Censorship on the individual and collective scale works in lockstep to preserve the possibility of pleasure [*jouissance*] – understood here as a form of destruction, inertia, collapse of the inside, and absence of internal limits. The subject learns to its own detriment to make use of what besieges it. Although overwhelming for subjectivities, censorship also offers tremendous advantages, since many of the barriers it imposes are perceived as fictive. While yielding to these barriers, the subject also spends a great deal of time devising schemes to bypass them unseen. Except that, on the visible stage, it must convey perfect obedience so that its internal *détournements* are not unmasked. Once beyond the barriers, the subject arrives at a site within its subjectivity where vastness and chaos, light and shadow, desires and taboos become indistinguishable. In this tragicomic drama, the subject strengthens to its detriment public censorship by becoming in the eyes of others a loyal defender of the established order. In this way, specific modalities of political power are embedded in each individual. As Mohammed Dib writes: "But you have been left with total freedom! Only the freedom of others is at stake."[13] *Détournement* thus disarms the subject's insurrection against censorship, as it moves from making compromises with censorship to becoming almost permanently compromised by it. The totalizing force is thus able to break the subject's necessary radicalism by guiding its existence.

In this dynamic, oppressive game between different forms of censorship, permission and prohibition, the real responsibilities of the subject (and of the political order) are clouded over as it struggles to orient its existence. Paradoxically, the process of internal emancipation doesn't lead to any sort of liberation. Instead, it breeds the feeling of fear, both of one's self and of the Other. The transgressions carried out in the private sphere bring some immediate satisfaction, but the brief trip taken ultimately leads back to where it all started. Liberation turns out to be a mere dream, one that is poorly understood by the political power and, worse still, by the subject itself. The subject is forever performing a balancing act between emancipation and imprisonment. Its performance is both discreet and innovative as it seeks another exit while still respecting the general choreography. But its innovations then become part of the main act, and the subject gets lost in its own duplicity.

Since the end of the 2000s, censorship in Algeria has grown more complex and subtle. Algerian literature today, for its part, has grown richer and more impactful thanks to the art of *détournement*. Censorship strives to create a unified public by homogenizing subjectivities. This isn't unique to Algeria. It takes place in other forms all over the world. A purely cultural view of Algerian society fails to note the role politics plays in this and the paradoxical workings of a stratified censorship, which has become almost a sensory experience in and of itself. The subject both fights and gives in to its antagonists (internal and external).

An exacerbated religious morality seems to be the vehicle of communication for these censors. It has invaded the various spheres of private and public life to such an extent that morality, religion, and culture appear to have merged. This invasion has affected the very means of co-existence as well as new areas of the subject and its institutions. It is most acutely felt where morality has replaced tradition, which once provided a fundamental source of social cohesion by establishing, reinforcing, and celebrating a common set of references. It enforces political taboos by excluding any form of alterity. Censorship maintains the status quo between the subject and the community, between the subject and the political order, and finally between the subject and the Other who lives within it. Obeying the censors offers the major advantage of appeasing interior conflict, but this conflict is consequential for subjectivity. Tired of fighting, the subject may prefer the advantages offered by remaining morally vigilant. Make no mistake, censorship corrupts from within, not unlike the corruption affecting the national economy.

Tradition, insofar as it brought together heaven and earth, the human and the transcendent, the visible and the invisible, curbed violence and tension and mediated between generations. Tradition often embraced conventional morality, but it was never reduced to it. We are witnessing today the transformation of tradition into religious morality. Previously, tradition was made up of a diverse set of religious and pagan practices which were inherited from a plurality of regions and linguistic cultures. Many distinct worlds fanned out across all of Algeria, each with a multi-faceted belief system and a rich array of practices. Religious rituals within these traditions were borrowed from the three principal monotheistic religions, especially Islam, and even today (although increasingly less often) one can find aspects that hark back to this hybridity.

For example, in the Chenoua region (in the Tipaza province), to celebrate Eid al-Adha, a small glass of fresh sheep's blood was to be drunk by the person who carried out the sacrifice immediately after the animal's throat had been cut, which recalls the symbolism of the blood of Christ in Christianity. Similarly, in the district of Djanet (Illizi province) in the Algerian desert, inhabitants of this region still recount the legend of the Jewish origins of the Sbiba festival, which is celebrated during the period of Ashura (a Muslim religious festival) and which commemorates the Jews' exodus from Egypt. Amin Zaoui's novel *Le Dernier Juif de Tamentit* (2012) evokes in a similar vein the long history of a Jewish tribe that lives in the Algerian desert. The story unfolds in the city of Tamentit, in the Touat region (Adrar province). Blending historical fact with legend in a narrative account, it displays the mosaic of region-specific traditions that span an immense country. "I like stories where different histories mix together, where there is an itch to stitch and unstitch," Zaoui writes. "I love confusion, the tangle of language! The joy of frenzy!"[14]

The power of the "language, religion, and politics" (LRP) bloc as revealed by clinical psychoanalysis

Islamists are not wrong when they claim Algerian traditions contain what they deem elements of impurity. Thus, they call for a veritable dismantling of tradition by eradicating the existing state of affairs. This attempt to dismantle tradition is nothing new; but it has been dramatically aided by capitalist, and its corollary, technocratic, discourse. Post-Independence Algeria has taken it upon itself to continue this dismantling. It accomplishes this by collapsing the distinctions between language, religion, and politics, which leads to a configuration I have designated as the "LRP."

It is worth pointing out how religion's spiritual, civil, and social function has been distorted [*détournée*]. Religion is now no more than an act of passing judgment between what is accepted (*el halal*) and what is taboo (*el haram*), between *yadjouz* (literally, "this passes") and *la-yadjouz* ("this doesn't pass"). It is difficult to fight this type of judgment, which is brandished like a weapon in all types of discourse. As for the private sphere, one's relation to self-judgment – in the sense of making demands – is defined by the need to obey, therefore pre-empting critique and resistance. As a result, when insurrection is visibly and openly waged, it will inevitably be violent. This is why the subject spares itself this violence by opting for private transgressions, which, without effecting any real change, are in reality just another form of moral obedience.

This moralization is meant to foster social cohesion and offer a shared set of references. But it also serves as a straitjacket for the subject, who can neither intervene nor resist. Each individual is thrust into his or her own solitude and left to devise spaces where culture and knowledge regulate internal (self) and external (social) relations. Reducing religion to moral conduct is a way of imposing uniformity, which is driven by a hatred of difference.

In *The Future of an Illusion* (1927), Freud explains that the civilizational process, which allows people to live together, is fragile. People are in need of culture to control their violent, anarchic impulses, but, at the same time, this domestication leads to many sacrifices: "[E]very individual is virtually an enemy of civilization, though civilization is supposed to be an object of universal human interest."[15] Freud adds: "Human creations are easily destroyed, and science and technology, which have built them up, can also be used for their annihilation."[16] The importance of the individual's ambiguous relation to culture cannot be overstated. Culture serves a need to create, then maintain, a subjective space that makes social cohesion possible. But this is countered by another force, a destructive one that pushes individuals to work against themselves internally by demolishing their progress toward civilization.

Each individual is therefore responsible for helping construct and maintain the civilizational process. Culture begins with the acquisition of language and speech.

This initiates a break with the organic state (or "animal" state, according to Freud). Viewed in this way, education, instruction, and knowledge work to enrich and strengthen the construct of culture, and, by extension, the construct of the human, by keeping the human's destructive instincts in check. But culture's hold on the subject is constantly threatened by interior and exterior forces. Today, the moralization of religion is the foundation of culture and knowledge in Algeria. No one living in, or visiting, Algeria fails to notice this. Moral pronouncements have risen to a fever pitch: everyone has his or her own interpretation of religious morality.

Patients are not immune from this moralization of religion. Their perceptions of the effects of analysis during treatment reinforce the existing state of censorship. For its part, susceptible to becoming yet another vehicle for extending the moral reach of the LRP, psychoanalysis faces its own troubles.

Psychoanalytic treatment exposes how censorship, taboos, and social conventions govern thought and speech. Analysis is perceived sometimes as a threat and sometimes as an ally in this struggle. Indifferent to conventions, it encourages private speech, serving the individual and yet allowing for the collective nature in each individual to be revealed. Speaking freely shows that censorship, although externally imposed, is aggressively internalized. Overcoming it is a personal matter. Reaching this point, patients still have a very long way to go. Will they recognize, and then refrain from, the role they play in abetting the empire of censorship? And if so, what to make of a freedom deprived of the social space where it can be exercised? Isn't it safer to keep up the minor transgressions in the dark, a practice much less costly and promising its own secret pleasures?

With psychoanalysis, patients first find themselves delightfully surprised by the suspension of moral judgment. But a reversal then occurs and over time things only get harder. It becomes hard to discern between what may be triggered by social taboos (charged with religious signifiers) and patients' own self-censorship. What is particularly striking is how the subject ends up fully and unquestioningly embracing the taboos it is besieged by. It hasn't lost its ability to think for itself, but this ability is increasingly placed out of its reach.

This relationship with censorship can be found in any patient, regardless of his or her language and the place of treatment. But usually, at some point during the treatment, the subject steps out from behind the taboos imposed from the outside (family, education, religion) and begins to question its own practice of self-censorship. In Algeria, reaching this point of dissociation has proven stubbornly difficult. Taboos continue to exercise control over thought. For this reason, the subject finds ways to hide behind its speech so as to avoid being caught off guard and fully exposed by religious morality and the reigning ideology. And yet it seeks treatment in order to find a self it may no longer recognize. But one mask only gives way to another in an endless cycle as the subject continues to prop up the censorship under which it is so clearly suffering.

Seeking cover within its own subjectivity, the subject gets lost amid its own disguises and fear sets in. Indeed, it grows fearful of its own movements in the silent darkness, afraid it may disappear into the blank space of speech. Nabile Farès evokes this state:

> Fear of oneself, fear of others. Fear of oneself: yes, as though haunted from within; haunted in the most visceral way, as though you could feel the brittle limit of your life, right there, inside your body; as though your body defined the limit, the limit of resistance and of duration; *as though you needed to learn how to hide your body, just like you learn to hide your feelings.*[17]

Farès reveals with this the secret of the body/feelings to be covered and hidden: the ferocity of fear and its effects. Indeed, this fear creates the many "masks": ideological, moral, political, and other markers of identity. Is censorship the control center of fear? Does one's fear correspond to the severity of censorship? And might censorship be responsible for the orchestrated confusion over who speaks, who imposes taboos, who thinks? Is the invisible force pulling the levers both a subjective *and* political matter? Fear surges forth near the dismissed zones of discourse and thought, as there are no support barriers there. As previously indicated, the subject's masks multiply in this invisible space. It seeks to find refuge for its most taboo thoughts so that they won't be confiscated from it. However, in the process of looking, it digs deep into the horror of the blank space.

Initially, the subject dons a whole series of masks needed for its acts of *détournement*. But then, little by little, it finds itself the victim of its own act. The more it loses itself in its roles, the closer it comes to the fear it strove to escape in its playacting. The whole affair unfolds outside of speech in the greatest secrecy. The scale of this "silent act," one that remains protected from onlookers, raises some pressing questions: to what extent is the subject's act a performance? Does this personal performance mirror the political dynamic of the larger public?

This sheds light on how the individual bolsters the LRP bloc by unwittingly performing its dictates. A growing religious morality suppresses differences in lifestyle, thought, and beliefs. It doesn't allow for any separation between the inside (one's superego) and moral principles. This lack of separation makes it hard to know precisely who – which superego – is speaking. Is it the subject or the voice of a community of believers converted to a new form of Islam spread from the Middle East? There is no room for this question in the subjective space of the patient. Both voices merge to form a single entity, resulting in an endless internal struggle. Treatment takes place amid this war with censorship, which, striving to contain fear, ends up making it spread more aggressively. In the land of the LRP, for both the patient and the analyst, it is hard to get over this embattled struggle.

The duplicity of subjects confronting censorship from the LRP

In Algeria, subjects need incredible strength and energy in order to keep dreams and desires within the realm of possibility. This can lead to new subjective discoveries that are all the more pleasurable for being hard to achieve. But the effort to get there can be overwhelming. Oftentimes the subject prefers to give up rather than let go of the social and imaginary taboos to which it clings.

If all desiring subjects (no matter where they are) confront the same problem, in Algeria the internal work performed by the very nature of the human psyche is crushed by the demands made by the larger public. In this case, the outside is no more than an eternally retreating space incapable of accommodating a troubled inside. Freud helps explain what a desiring subject endures when struggling with morality. In Algeria, morality has the final word and the public has become the one who delivers it. "Anyone thus forced to react continually to precepts that are not the expressions of his impulses," writes Freud, "lives, psychologically speaking, above his means, and may be objectively described as a hypocrite, whether he is clearly conscious of this or not."[18]

Desire must be discreet and cunning in order to counter the subject's morality, which is reinforced and amplified by the larger public, morality's faithful guardian.

Confusion becomes the subject's best tactic for creating a secret, off-stage site within its subjectivity and for deceiving fear (from within and without). These ploys occur on a daily basis, and one of the most noteworthy among them concerns conjugal schemes, especially for gay men and women.

Gay women cunningly make use of the social division of gender in public space to make everyone believe they're merely living with a female friend rather than living as a couple. As opposed to gay men, very little acting is involved. Indeed, until recently, unmarried women rarely left the parental abode to live alone or with another individual (male or female), as this choice was viewed as a sign of sexual promiscuity. These women speak openly about their *housemates* rather than their *partners*. This clever strategy is a perfect illustration of how performance is used as a form of *détournement*. These women use the division between sexes and the traditional practice of placing them within one homogeneous group to their own advantage. Indeed, forced to be together, it requires almost no extra effort to live out their sexual and romantic lives. And so, unseen by the censors, they satisfy their desires under a regime where homosexuality is banned while following the law established by men to keep the sexes apart. Paradoxically, they benefit from respecting this division of the sexes, which also makes it easy for them to engage in sex outside of marriage. Staying quiet and out of the spotlight, they avoid making themselves targets for exclusion and repression. It requires minimal effort on their part. And, for this same reason, this practice of *détournement* does nothing to change the taboos in place. The public display of obedience persists and, as a result, *difference* is never witnessed since everything happens behind the scenes.

The practice of *détournement* opens up a whole field of desires and fantasies, but it does nothing to challenge censorship and other taboos, which remain in full force. This subversive, and at times transgressive, tactic is subtle, as it outwardly conforms with censorship. Why flout the censors when you can profit from them at little cost to satisfy your desires? This logic can be found operating on many levels. At its best, it can subvert private life by bringing to light valuable discoveries on the social stage. But, more often than not, as the endless performances of *détournement* remain invisible, they are robbed of their subversive potential. The reign of secrecy is upheld by all as the (silent) path of salvation against a deafening censorship. A potentially subversive act that can lead to change and transformation is thus disabled. Right where an irreversible break should have occurred we have instead invisible performances played out in the dark. *Détournement* teams up with subversion only to cancel each other out.

Whereas *détournement* upholds the established order, allowing for only a small number of desires to be satisfied, subversion overthrows it. In other words, subversion creates staggeringly new signifiers that bring the endless performances to a stop. *Détournement*, for its part, does just the opposite, churning out performance after performance with no end in sight. This clearly prevents a new subject position from emerging, which could lead to an irreversible emancipation from censorship. Instead, there is an endless game of hide-and-seek going on in private and social life. Difficult to identify and delimit, censorship creates a need for transgression. This is felt by the subject every day, but it also signals a more serious transgression: material and moral corruption, including the deliberate distortion [*détournement*] of existing laws (see chapter 7). *Détournement* works with and against disruption. The fact that citizens perceive the law as inoperative is a case in point. Each individual is left to fend for him- or herself. The subject is at the mercy of inexplicable taboos designed to serve the interests of some individuals as well as amoral and arbitrary laws imposed by the religious sphere.

Through the practice of *détournement*, minor transgressions that allow some small space for desire in the private sphere (no simple feat in itself) are not considered compromising. These transgressions follow a well-beaten path. However, recognizing these and expressing them in the social arena comes with great risk, a risk felt both in the imaginary and in the realm of the real. In light of this, as a process of exposing and recognizing deeply buried desires, fantasies, and dreams, psychoanalytic treatment represents a real danger and poses a unique threat. Fear spreads uncontrollably, reinforcing taboos and creating a permanent duplicity, as noted by Mohammed Dib: "Duplicity is rooted to our most private selves. Ridding ourselves of it would require completely disassembling, and then reassembling, the self."[19] In other words, the subject prefers the comfortable satisfaction afforded by voicing grievances over the violence of assuming responsibility for one's decisions. And for good reason. In the social arena, it would be assuming that responsibility all alone.

If religion is an astonishing vehicle for morality, the specificity of the LRP bloc lies in its unique melding of the two. The two terms are *almost* interchangeable. The strong collusion between morality and religion serves multiple, and paradoxical, interests. Each individual works on behalf of the censors he or she is besieged by like an agent of a larger system. For this to work, the subject must benefit in some ways from the same situation that is responsible for its downfall. Putting this experience into words is a telling acknowledgment that exposes the subject and forces it to abandon irrevocably its secrecy. Insofar as it can break the social pact, whose terms the subject has implicitly agreed to, exposure of this sort raises serious threats and risks.

Abandoned citizenship and speech acts

The obstacle that every speaking being encounters is dealt with by religion rather than individual speaking beings themselves. This reflects a citizenship that, while still trying to find its bearings, is repeatedly undermined. Given the floundering state of institutions after years of internal war, the faithful are easily replacing citizens. This replacement is occurring at a time when various forms of corruption are staging countless performances designed to abuse [*détourner*] laws and other regulations.

Individuals dealing with institutions experience this on a daily basis. It can even be seen in their own speech as they distance themselves further and further from the private sphere. There remains a *presumed* citizenship, since that was the goal and purported achievement of Independence, but it can't be exercised. To use the words of Frantz Fanon (1925–61), it is but a "hollow title."[20] Fanon, a psychiatrist and brilliant thinker, had already warned us well before Independence of the dangers of a hollowed-out citizenship and its relegation to a meaningless title stripped of its function.

The anthropologist Mohamed Mebtoul, whose research examines the relationship between individuals and institutions in Algeria, underscores the widespread feeling among individuals that not only is little offered them in terms of help, but that also their needs are routinely unmet or neglected altogether. Mebtoul relates this lack of care to the question of citizenship. He speaks of a "missing citizenship," explaining that "a social system that aggressively opposes free and open debate with its own line of thinking and imposes its own suspect ideology as a means of indoctrination all but forbids citizenship."[21] He shares what a young despondent individual told him during an interview: "The only support we find comes from the wall." This young individual is a "*hittiste*," literally someone "held by the wall." This is how young people without jobs are referred to (and refer to themselves). This bodily presence lining walls throughout big cities is less of a problem today. Resourcefulness (*chtara* and *gfasa*), the art of *détournement*, and economic hardship have forced this population to leave their position along the walls in major cities. Mebtoul concludes his study by claiming that "the zones of precarity are a product of a social modality that thrives on organizational confusion."[22] A hollow citizenship invites an endless and unlimited use of *détournement* while no longer being able to identify the stakes or actors at play. This perfectly illustrates how systemic corruption becomes law.

This sense of confusion spreads aggressively, affecting all levels of social organization. It has also been dramatically exacerbated by the Internal War, which has weaponized widely circulating signifiers and common war strategies. The cult of invisibility, along with its corollary, an excess of visibility, only accentuates this confusion. Among this hazy landscape, the way in is often confused with the way out. Is it a matter of finding a "way in" to liberate the individual in favor of a responsible subject or rather of seeking a way out of a particular narrative of liberation that, in its long and dramatic unfolding, has been taken for granted and gone entirely unquestioned?

In psychoanalysis, interpretation, which strives to offer a new subjective experience, is hindered by this "missing citizenship." The history of Algerian Independence is re-enacted by depriving individuals of an inner revolution. The subject strives toward inner freedom, but never acts on it. Unbeknownst to it, the power of servitude holds it in place. These paradoxes plague psychoanalytic treatment.

Another inventive practice adopted by the subject confronting the LRP consists of creating its own internal separations by purging itself of the moralization of religion. It is hard to find a term for this singular discovery of a neutral, mediating space: can we say in this case that this is a form of private secularism?

Men and women try to carve out a neutral space as far from the LRP as possible. They accomplish this with language, knowledge, and culture. This form of secularism that begins as a conceptual exercise paradoxically takes shape only in the private sphere – it is entirely absent from public space. One might well object that the expression "private secularism" is an oxymoron. But this interior state is clearly observed among a certain number of individuals within the city. These individuals find themselves compelled to invent new modes of navigation in order to steer toward a wide-open horizon. Very often they pay for this with a staggering sense of solitude, which is difficult to cope with in the long run. In other words, some don't ask for permission and live out their desires in defiance of the law. The realization of this fundamental transgression is not without risk. Many intellectuals have been murdered seemingly over this. More precisely, these political assassinations took advantage of the freedom of speech demanded by these intellectuals in order to sow doubt about who was behind them. The message couldn't be clearer, spreading fear and distorting understanding. Self-censorship targets free thinking, difference, and unconventionality. This type of censorship is extremely refined. Its introjection has been so successful that its centers of operation (linguistic, political, religious, etc.) have become entirely unidentifiable. In contrast, in "regular" totalitarian states, censorship can be identified in its grand gestures such that action can be taken to prevent it from invading the private sphere. With the LRP bloc afflicting Algerian politics, the forms and levels of censorship at work in the subject become blurred. Thought and speech are undermined in favor of a literal form of discourse and language that are stripped of nuance and ambivalence. The only option left is "to speak by withdrawing from one's self, to speak in isolation from one's voice."[23]

To understand and analyze the history of this fear on the level of individual subjectivity and in the context of widespread social decline requires a deeper look at the facts and effects of History. The traces of colonization are drowned out by a noisy conflict that leaves no mark. We will try to penetrate beyond the noise that rings like a case of tinnitus, troubling the "sleep of the just"[24] in a crumbling world, in order to hear the faint voices reverberating without content, right there, at the beating heart of history, where writing fails to materialize.

2
Colonial Rupture

> Colonial discourse is always pointing at the colonized to announce its own downfall.
>
> Jacques Hassoun, 1999[1]

> Nothing affects the heavy anger of the oppressed creature; he doesn't count the years; he doesn't distinguish men, or roads; for him there is only one road; the Roman road; the one that leads to the river, to rest, to death.
>
> Kateb Yacine, 1956[2]

For centuries, Algeria has been the site of many conquests and of battles fought against the Carthaginians, Romans, Vandals, Byzantines, Arabs, Turks, and the French. Each of these upheavals has had an impact on the habitus and languages of this land's inhabitants for many generations. But the impact varies with each period, and no continuity emerges from the shifts occasioned by successive invasions. Speaking to this point, the poet Kateb Yacine (1929–89) writes that: "In their country neither the Numidians nor the Barbary pirates conceived in peace. They leave it to us virgin in a hostile desert, while the colonists follow one another, pretenders without law and without love."[3]

One of the defining features of France's conquest of Algeria was its baseless claim that this land had no previous history or culture; that it was virgin land awaiting its conqueror, which gave rise to a very particular phenomenon: the impression – in its first meaning, like the impression of ink on paper – among colonizers, "*indigènes*," and all those involved in this process, of a blank space in history. The rich history embodied in the land's languages, myths, poetry, and traditions had been abandoned. Designating the autochthonous population "*indigènes*" attests to this fantasy of a people without history – the founding myth of coloniality. Neglecting the series of conquests that preceded its own, France saw its colonization as unprecedented. The effort to erase languages and history is a unique trait of French colonialism in Algeria. The protectorates of Morocco and Tunisia, which were imposed at the beginning of the twentieth century, never experienced this effort by France to eradicate an "indigenous" past. Nor did the colonial practices of Britain take this approach of systematic erasure – which may explain the lasting co-existence of many languages and belief systems in certain countries such as India.

French colonization, understood as a system of indiscriminate destruction and occupation, operated on many levels – affecting both the individual and the larger public – and in a variety of ways: through language, history, faith, and tradition. This violence has left irreparable marks on bodies both real and symbolic (such as language), but these marks aren't always visible. Indeed, a principal feature of colonial destruction is its deliberate confusion of *before* and *after*, which blurs the lines of continuity and discontinuity and makes it exceedingly hard to separate causes from consequences, beginnings from after-effects. Nor is it easy to draw on history to assign full responsibility. In order to accomplish that, a careful and dispassionate analysis is needed, one that would invite the writing of history from a plurality of individual perspectives.

The colony: the rogue child of the Enlightenment

It is important to remember the place the colonization of Algeria occupied in France's nation-building project in the nineteenth century. Since the emergence of the First Republic (1792–1804), political power in France was fighting the specter of the *Ancien Régime*. The French Revolution's radical break with absolutism caused certain areas of the previous regime to go dormant and await their return. Despite the evolution of French politics and its liberations by the Republic (in 1848, then in 1870), Algeria, conquered by a "constitutional monarchy," would continue to suffer under the violence of a monarchical and totalitarian power.

In July 1830, while Algiers was being seized by the French military, a "parliamentary and liberal monarchy" was established in France to take the place of an absolute monarchy.[4] In France during this period, the Republic remained a threat to be warded off. After the Third Republic (1870), there seemed to be a clean break between the two systems of governance. However, what was suppressed in the metropole by the Republic was unleashed in the colonies: hatred for republican principles would remain a constant in colonial policy. A repressed monarchy reasserted itself on another land, where it reigned with unchecked power throughout the colonization of Algeria.

Observing France's evolution from the vantage point of the colonization of Algeria in 1830 reveals the unbreakable – though invisible – ties between the Republic and its underside: totalitarian power. Many historians have argued that colonization was seen primarily as serving domestic affairs. It was a strategy to placate and control internal tensions while solutions were sought to the crises plaguing France. Ambitious economic, social, and political goals were needed to generate enthusiasm and establish an economic powerhouse. Kateb Yacine expressed this in the following way: "[T]he conquest was a necessary evil, a painful graft promising growth for the nation's tree slashed by the foreign axe; like the Turks, the Romans and the Arabs, the French could only take root here, hostages of the fatherland-in-gestation whose favors they quarreled over."[5]

In the French political arena, the turbulent and messy transition from a singular (monarchical) power to a plural (democratic) one triggered a violent response that repeatedly targeted the colonies. Indeed, this unresolved and unceasing violence was a major export of France. Sometimes this violence made its way back to the metropole, as in 1848, when barricades were set up across Paris on the eve of the Second Republic. Protesters were called "Bedouins" by some political officials and violently attacked, often by the same ruthless officers who led the conquest of Algeria. Let's recall the famous words uttered by Louis-Napoléon Bonaparte in October 1852 during his speech in Bordeaux: "Empire means peace." In Algeria, the "*indigènes*" were not considered "French", but "French subjects." "French subjects" unequivocally recalls "the King's subjects." A clear shift in the signifier "subjugation," but this time for others, the "*indigènes*." Colonial policies therefore actively commemorated the Republic's pre-history (absolute monarchy) while France continued to multiply its colonial conquests almost a hundred years after the Revolution.[6]

The colonized lands only make the artificial division between monarchy and republic all the more apparent. The creation of colonies doesn't represent a contradiction within the Republic, nor is it at odds with it: coloniality is what naturally emerges from the Republic, like terror or tyranny, rarely failing to pay tribute to the monarchy that begot it.

Algeria thus reveals both sides of the Republic. That both sides would come together in Algeria seems paradoxical, a fact that is often overshadowed by other towering divisions. This concerns the treatment of individuals whose status places them on the margins of the Republic (Jewish/Muslim "*indigènes*" vs. "Europeans"). Civilizing a savage population (which is known as "*la mission civilisatrice*") was an integral part the French Republic's political agenda of 1870. The symbolic systems of these peoples were deemed non-existent. The Republic was exporting the flipside of its Constitution (the Declaration of the Rights of Man and of the Citizen). Thus, hatred for the Republic dictates how the colony is governed, a colony thrust into a night devoid of sleep and dreams. Does the Constitution of the Republic contain an unspeakably dark side that remains at play?

The period's politics of conquest drew on an international and universalizing power. Wealth during this burgeoning period of capitalism was measured by territorial acquisitions. Slavery and colonization provided the engine for this accumulation of land and wealth. In the French Republic, the trafficking of human beings took place in far-off possessions, which exclusively served the economic interests of the metropole.

When French troops debarked at the port of Sidi-Ferruch in 1830, Algeria was part of the Ottoman Empire and was governed by a Bey. The purpose of the trip was supposedly to settle an unpaid debt incurred in Paris a few years prior. This episode, which serves as a sort of foundation myth, occurred in 1827 and is known as the "*coup d'évetail*," or the "fan affair." The story goes that France was to pay the Bey of Algiers for a large purchase of wheat made in the early 1800s. The Bey, angered by the failure to pay this debt after his insistent requests, and by degrading comments made about Islam, is said to have struck his French counterparts with the fan he was holding. Officials in France claimed the conquest of 1830 was a response to this *offense* from three years prior. But historians are clear on this point: this was far from a simple response. Given the scale of the military operation, a significant amount of time was clearly spent devising a careful plan. The *offense* at the heart of this story would become the standard justification for the colonial project, endlessly repeated at the time and still unquestioned by some today. *Offense*, a signifier that trumps all others and the cornerstone of countless arguments, is imagined to distribute hate equally between these two countries, which finds expression, both implicitly and explicitly, in their respective political discourses.

The case of an unpaid debt for a purchase of wheat signals at the very outset what would define the colonial project, namely, appropriation for some and expropriation for others. Perhaps more important still was the commodification of goods and human beings that was now occurring outside the traditional realm of spoken agreements. This commodification would quickly transform the social contract. Previous signs of wealth, including agricultural abundance, would soon be replaced by a monetized system.

In his literary works, Kateb Yacine observes that the colonizers' arrival seemed initially to obey the logic of exchange, but that it proved to be a violent seizure of land and men. Merciless appropriation came to be a recurring phenomenon throughout colonization, which, forgoing negotiations, left one party in a world of destitution and destruction with a serious loss of material goods and, it bears emphasizing, immaterial ones too, such as cultural and symbolic knowledge. In contrast, the Ottoman occupation, for all its abuses, did not destroy the existing tribal structures in Algeria. Modes of exchange and manners of identifying kinship, as well as the languages spoken, survived unscathed. And let's not forget that Islam – as a system of beliefs and a philosophy underlying social cohesion – was shared between a majority of the autochthonous peoples within the Empire and the lands under its control. However, this did little to prevent various tribes within the country from rebelling against an occupation that, though without abolishing their culture, was imposed against their will.

Colonialism's destruction of social cohesion

Between 1830 and 1847, a violent war of resistance was waged against France. These war years pitted various tribes against the new settlers and led to a significant loss of human life: according to historians, nearly a third of the population was wiped out by human massacres or famine.[7] The aim was to kill as much of the autochthonous population as possible and to deploy a permanent reign of terror in order to "squeeze out" the Arab population, to use the appalling expression of the French philosopher Alexis de Tocqueville (1805–59).[8]

In his seminal work *L'Honneur de Saint-Arnaud* (1993), François Maspero (1932–2015) shares the accounts of those who witnessed the invasion and the employment of *enfumades*. This practice, which forced people into caves that were then filled with smoke in order to suffocate them, wiped out entire tribes. Many were ordered by Colonel Edmond Pellissier de Reynaud (1798–1858), who wrote, for example, that "Everything that lived was condemned to death. ... Returning from this expedition, many of our soldiers carried the heads of their enemies impaled on their lances, one of which, I was told, was served at a horrible feast."[9] Several stories from this period show how the "*indigènes*" were treated like beasts. This is in line with a statement from Tocqueville, written in 1847, where he claims that "it seems as though, seeing what is going on in the world, the European is to the men of other races what man himself is to animals."[10]

These wars also led to the looting of resources and the destruction of the elites. The collapse of traditional structures and the aggravation of poverty also date from this period. Fratricidal tensions between tribes at that time were intentionally deepened to create more conflict.[11] The conquered land wasn't a single nation. It was governed by tribes and religious groups, which differed from region to region. In fact, some parts of the land weren't conquered until much later. France's invasion didn't advance at a steady pace, as tribal leaders – the most famous among them being the Emir Abdelkader (1808–83) and the Sheik Mohammed El Mokrani (1815–71) – staged their own revolts. The French army's chief strategy became one founded on disappearance: making lands, goods, and bodies disappear *en masse*. Historians speak of a "war of extermination," a "total war," and "mass murders" when addressing this moment in history.[12]

Kateb Yacine attributes to this period the emergence of the widespread feeling of offense among the Algerian population, a feeling that, passed down from generation to generation, grew stronger throughout colonization and continued to persist well after its end. He explains how the arrival of money used to purchase land (the rare land that had not already been expropriated by force) led to a transition from a solid materiality to a liquid one:

> The chiefs of tribal Algeria, the ones who had access to the treasures, the guardians of tradition, were mostly killed or dispossessed during those sixteen years of bloody battles, but their sons found themselves facing an unhoped-for disaster: ruined by the defeat, expropriated and humiliated, but retaining their opportunities, protected by their new masters, rich with *the money their fathers had never turned into liquid assets* and which the colonists who had just acquired their lands offered them in compensation, they had no idea of its value, just as they no longer knew, facing the changes brought by the conquest, how to evaluate the treasures saved from the pillage. ... The fathers slaughtered on Abd-el-Kader's raids [...] had never taken inventories; and the sons of the vanquished chiefs discovered they were rich in money and jewels, but frustrated; *certainly they felt the offense*, yet in their retreats they had not kept the taste for the battle that was denied them.[13]

This describes how social structures founded on spoken promises and a codified system of exchange functioned, and, more importantly, how they were being destroyed.

The exchange of one's word, which was treated as though legally binding, was replaced by violence, and this shift had enormous consequences for the social structures underlying Algerian society at that time. Even today this is registered in the grievance of patients who lament the refusal to acknowledge spoken agreements, whose social function has all but vanished. This is expressed in Arabic as *mabkatch el kelma* – literally, "one word doesn't remain." In traditional societies, exchanges were carried out with a simple "spoken agreement," referred to as *el kelma*. This served at once as a transaction, a social pact, and a signature.

An older generation frequently comments on the steadily declining status of spoken agreements among today's youth. Speech, they remark, has been stripped of the agency it once possessed. Elders suggest that something must have led to this dismissal of speech. But they struggle to identify the causes of the word's disappearance. Has speech transformed from a solid state into a liquid one? What accounts for this liquefaction?

A colonial republic divided, or the "duty to civilize [the] barbarians"

To address these questions, it must be recalled that, throughout the world, nineteenth-century capitalist expansion emerged from the abolition of the social pact that had previously governed each society. In colonized Algeria, using violence to commodify goods and humans was a large-scale experiment – one that would become widespread throughout the world in order to make exchanges uniform. Criminal violence would become associated with the treatment of humans as objects in a so-called "modern" economy. The equivalency drawn between goods and people, firmly established today throughout the world with its totalizing approach, was a product of the colonial enterprise. Racism and segregation are the driving forces of modern capitalism, and they would continue to provide the theoretical basis of colonization. This is where the cultural mission to civilize others irreversibly falls apart, as the anthropological structures that held people together came under attack. The historian Benjamin Stora speaks of this "terribly cruel" motive for conquest, adding that "what makes the colonial world incomprehensible, to a certain extent, is its erasure of its origin in hatred and warfare."[14] This erasure is tantamount to an attack on history, on the myths and anthropological foundations of the colonized populations.

When colonial expansion became a public debate in the 1880s, the French political world found itself divided. Underlying this debate was the thorny question of the republican pact, which colonial expansion calls into question. This pitted those – such as Georges Clemenceau – who supported republican principles against the specter of monarchy, which was rearing its head in the imperial policies of the time, against those who had always despised these republican principles.[15] The "march of civilization against barbarism," as Victor Hugo said in 1841, defined the colonial ideology.[16] Paradoxically, this ideology became an extension of the Declaration of the Rights of Man and of the Citizen, even as it undermined the very foundation of traditional societies. This same sentiment is also found in Tocqueville, attesting to the great rift dividing the discourse of this period:

> Muslim society in Africa was not uncivilized; it was merely a backward and imperfect civilization. There existed within it a large

> number of pious foundations, whose object was to provide for the needs of charity or for public instruction. We laid our hands on these revenues everywhere, partly diverting them from their former uses; we reduced the charitable establishments and let the schools decay, we disbanded the seminaries. Around us knowledge has been extinguished, and recruitment of men of religion and men of law has ceased; that is to say, we have made Muslim society much more miserable, more disordered, more ignorant, and more barbarous than it had been before knowing us.[17]

Despite this clear-eyed account, Tocqueville deemed colonization a necessary pursuit, even advocating for a complete "Frenchification" of the land – France being in his eyes the model civilization – while allowing them to keep "the yatagan and the baton."[18]

The ideology of the "clash of civilizations" was already beginning to take shape at that time. Except that in this case, the narrative it pushed was one of the civilized against the savage – let's recall that Berber (*Amazigh*) means both "free man" and "he who is not Roman." This is behind the statement that Jules Ferry made to the Constitutional Chamber in 1885: "If we have the right to invade their land, it is because it is our duty to civilize them. ... We must not treat them as equals, but rather remind ourselves that, as a superior race, we are their conquerors." The deputy Jules Maigne fired back in the middle of parliamentary debate: "How dare you say that in the country that founded the rights of man!"[19] This expression of the Republic's underside speaks to the rift dividing French politics at that time. Which explains why two radically opposed statements can be made in the same debate, even in the same speech of a single person, without appearing contradictory. Note that in 1885 parliament voted in favor of colonial expansion by a margin of one vote. A stark divide pitted the deputies who defended the rights of man against their pro-colonial adversaries, some of whom were calling for a "war of extermination" (eradicating an entire civilian population was even being discussed among anthropologists).[20]

As a result, throughout the colonial period, the entire population of "*indigènes*" faced severe forms of repression on a daily basis, including violence and humiliation. Torture was being practiced well before the start of the War of Liberation. Bodies were constantly being mutilated and put on display. The reign of terror long practiced by the colonial administration targeted the bodies of "*indigènes*" and impacted generations of men and women. The widespread practice of beheading and removing the ears of living and dead bodies,[21] which were sold in markets at 10 francs a pair, still haunts Algerians – Algerian Arabic has preserved the memory of this practice, as when someone who has been threatened by someone else says: *wach y dirli, y nahi li wadni*? ("What's he going to do, cut off my ear?").

The violence of colonial repression perhaps reached a new high during the 1871 uprising led by El Mokrani, which stretched from Kabylie to Constantinois. The large-scale massacres, forced labor, and seizure of the most fertile land caused unprecedented damage and cost the lives of a staggering number of people. The first wave of deportations to New Caledonia and Cayenne began at this time. The deported would go forever missing, banished from their land or the memory of the living.[22] Almost a century later, this savage practice of deportation and the "squeezing out" of the population was expanded by establishing internment camps in several regions throughout Algeria.[23] Add to this desolate picture hunger, epidemics, and, above all, the deep-seated feeling that human life has no value.[24] This last point endured more than any other, even well after Independence. Feeling spurned by the world and at the mercy of unpredictable violence delivers a blow from which it is hard to recover. Humans reduced to nameless bodies are held in constant suspicion on religious and racial grounds.

Not to mention legal ones. Serving at times as the Republic's (reversible) underside, at others its hollow center (a bottomless site of repression), the colony became much more clearly defined beginning in 1881 with the implementation of a *code de l'indigénat*, or a set of laws defining the status of the "*indigènes*" in Algeria.[25] The law dividing "French citizens" from "French Muslims" cast the autochthonous population as legal exceptions: the law and republican values didn't apply to them. This exceptional status would last through the entire period of colonization. French nationality was reserved for a select group of "*indigènes*," most frequently French-speaking individuals who had strayed from their social traditions. Nationality was therefore granted to individuals who were seen to be linguistically and culturally close to France, and, more important still, who were seen as having severed ties with the language and traditions they were associated with.

The shifting limits of legal exclusion culminated in the Vichy regime's revocation of French citizenship for Algerian Jews, whose presence dated back centuries. Speaking Tamazight (Berber),

Judeo-Spanish (Ladino), or Arabic, Algerian Jews were an integral part of traditional societies throughout the country (including the arid zones). In 1870, the Crémieux decree had granted French nationality and its accompanying rights to all those considered at the time as "Jewish *indigènes*" or "Israelites." Under Pétain, this decree was suspended on October 7, 1940. This barred Jews from practicing medicine and law and it also prevented them from teaching. Jewish children were also banned from public schools (the Jewish population then organized to establish schools with Jewish teachers). Algerian Jews would regain French nationality and citizenship on October 22, 1943. Throughout the colonial period, Algerian Jews thus clearly suffered under the whims of the Republic, welcomed into it only to be pushed back out on three separate occasions.

This history attests to the strong and active ties between the Republic and religion: citizenship was linked in practice to a single religion, Christianity, at the exclusion of all others. Following this logic, rights are determined by religion, which defines the limits of belonging both from within the Republic and outside it. This is why the 1905 law defining republican secularism didn't apply to the colonial project – this law excluded the "French departments of Algeria" – as the reigning ideology viewed Islam as incompatible with the French Republic. History seems to be repeating itself, as the term "French Muslims" has resurfaced today amid yet another debate on the compatibility between Islam and the French Republic. Let's recall that, after 1945, the "*indigènes*" or "French Muslims" depended on a separate voting body, the "second-tier electoral college," where a single "European" voice equaled eight "*indigène*" voices.

The War of Liberation demonstrated a clear refusal on the part of the "*indigènes*" to be reduced to the status of valueless objects. They were fighting against the same forces that excluded them from history, those that dispossessed them of their subjecthood by attacking their languages and the system of references that bound together the living and the dead. From a clinical perspective, the immensity and duration of destruction robs the subject of the possibility of feeling alive and of recognizing itself as "one" among others. The process of self-expulsion triggered by a vanished sense of belonging to human civilization creates over time a fractured life. It works by devaluing and excluding one's connections to lived experience. As Nabile Farès has expressed it, a sense of a "discounted life" dominates these "malnourished" sites: "Just try to understand a world that wants nothing of you. Just try."[26]

Interiorizing the experience of expulsion and being haunted by disappearance are cruel effects of the colonial machine. I will come back to how disappearance is at play on many levels. Those who go through this and survive are forever trapped in constant physical (rather than psychic) pain. It is from this site of beaten and mutilated bodies that Algerian Francophone literature is born.

1945: a literature of refusal is born

The site of speech's ruin calls forth a writing that bears traces of bodily disfigurement. Violence suffered by dismembered bodies and somatic feelings and metaphors permeate this literature. This writing aims to carve out a space for healing and repair, and for this reason it is written in French, as though it were necessary to take to – and take on – the tree of this language, to let oneself be soothed by its sprawling branches rather than by its hollow roots. The vast majority of the first Algerian writers were Francophone owing to their education. But there is something else behind the desire to submit French to poetic reworkings: it is a matter of seeking another form of pleasure in this language in order to gradually distance oneself from the harsh reality of the crimes committed in its name.

Colonial trauma is at the core of Nabile Farès' poetry. In his writing appears a body torn to pieces whose cries are forever lost in the vastness of an inhospitable world:

> While, buried under a few plots of filthy soil or disintegrated with quicklime in caves or hole of the local police station, I will bump into (yes, bump into) my own parts, and then search for my arms and legs, strewn about the city, the sea, each star, each movement of water, each speck of dust burning your eyes, your eyelids burnt by the wind; each word that, Yes, my nerves swelling the sky like branches, each word that I try to utter about this senseless search for my self, for a self scattered in a world set aflame.[27]

An amputated, "fractured" body pleads on behalf of its lost body parts in this literary work. The diminished, shattered body reflects the major impact colonial violence has had on subjectivities. Algerian writers make this impact visible. The pain is that of a missing body part seeking in vain to recover its lost wholeness. Discourse is also impacted, as seen by its obsession with war's persistence. Is this all just a natural byproduct of war itself?

Under colonialism, there is a distinct accounting of the living and the dead depending on whether "*indigènes*" or citizens are concerned. For Algerians, dispossessed of names and burials, the countless disappearances were anonymous and therefore incalculable. For the French, each death of their own was treated as an offense to the Republic and the values of its civilization. Beginning in 1945, there was both a mounting feeling of offense and a growing awareness of France's appalling brutality. This turning point is foregrounded in Kateb Yacine's *Nedjma*, where the war is depicted as both a "crime" and an "assassination of injustice": "Can crime alone assassinate injustice? Mother, I'm dehumanizing myself, turning into a leper-house, an abattoir! What's to be done with your blood, madwoman, whom can I take vengeance on for you?"[28]

This is a significant passage as it functions on two levels: historical and personal. It takes into account how History is formed through the history of the subject within a single temporality. This collage evokes the devastation of trauma and the impersonality of identification, which, whether it be recognized or not, is a straight-faced, one-directional process of History. A glide path joins (individual) history and (collective) History. The literary text serves this process by acting as a space for narrative, inscription, and, finally, metaphor.

On May 8, 1945, France celebrated its victory over the Nazis, a victory to which numerous Algerian "French Muslims" (*indigènes*) contributed, not to mention many other Africans, many of whom died in service of the French nation.[29] At the same time, vast demonstrations were held in the eastern provinces of Sétif, Guelma, and Constantine by "French Muslims" who were demanding better treatment and pushing for civil rights for all. The military's response was brutal, indiscriminately striking and killing protesters and civilians, women and children – Algerian authorities estimate today that 45,000 people were killed; French historians, for their part, put the number between 15,000 and 20,000.[30] The contrast between these two events – victory against fascism and the unleashing of another form of ruthlessness – which were unfolding at the same time within the same territory is staggering. Once again, the colony's unusual status as both belonging to, and defined against, the Republic is brought to light. A festive celebration against savagery in one place becomes, in another, a humanitarian crisis, leading only to more massacres. An irony of history: the African *tirailleurs* called on to carry out this massacre were the same soldiers who fought alongside the Algerians during World War II. This moment defines a definitive break, repressed by some (the colonizers), unforgotten and vivid for others (the "*indigènes*").

The massacres of May and June 1945 marked a turning point. After this moment, nothing could hold back an uprising that was long in the making. The uprising led to an outbreak of war, pitting an army against a population that would soon become targeted on a large scale – massacred, raided, interned in camps. People disappeared, bodies were mutilated and dismembered, with body parts of the living and the dead put on public display; torture became a common and frequent practice, increasingly so after France sought retribution for its humiliating defeat in Dien Bien Phu in 1954. From one "*indigène*" to the next. The Algerian writer Jean El Mouhoub Amrouche (1906–62) drew this terrible parallel:

> What was new and specific to Nazism – which must be seen as the outer limit and essence of colonialism – wasn't a widespread and quantifiable crime, nor the appalling notion of waging total war against an enemy, which is in the end all too human; rather it is the particular quality of a crime that posits that, under the law itself, anything goes, as long as it is directed against a human who has been previously excluded from humanity and stripped of his human quality.[31]

Kateb Yacine speaks of this event from 1945 on a more personal level. As a teenager living in the Constantinois region at the time, he was arrested and tortured. Following the disappearance of her only child, his mother went "crazy" and would never recover. Nedjma, the central figure of the eponymous novel published in 1956, is the name and story of a forbidden love, unfolding in a self-enclosed community where crime and incest fuel a fascination among men for a body deemed off-limits.

Nedjma: an esthetic of colonial destruction?

In the novel, Nedjma is a ravishing woman whose charm casts a spell over all her suitors. But she can only marry her brother. Both she and her brother are unaware that they are bound together by the crime of incest. The daughter of a mother who is a French Jew and a father who is a "French Muslim" (Arab), she was born from an adulterous and illicit affair. In her beauty, Nedjma embodies a symbolic outlaw and, as a result, leads to the downfall of anyone who gets too close. In the novel, her celestial name – "star" in Arabic – designates the downfallen being. There is something irremediable in this: can she rid herself of an identification marked by crime and incest? Breaking away from the infamy provoked by the dismissal of law proves difficult, if not impossible. All are blinded and intrigued by Nedjma.

External violence is immediately internalized. The mixing of inside and out becomes so prevalent that even the barriers dividing the living and the dead are no longer impenetrable: "Jugurtha sings what he feels when he examines himself," writes Jean El Mouhoub Amrouche. "Like Narcissus at the fountain, his cry expresses the eternal despair of a man abandoned by all, the plaything of an overwhelming, omnipotent force. This force isn't entirely external; its formidable power, he has no doubts, *comes from within*, and, no matter what he does, it is driving him to his own inexorable end."[32]

Coloniality's dismissal of law is featured in the literary texts of many Algerian Francophone writers by the blank spaces and gaps that interrupt the story. The text is marked by, and reveals, this loose relation to law. It is described in the trajectory of the characters and punctuates the writing itself (through blank space, abrupt halts, frantic punctuation). This goes beyond the notion of *translating* an emotional state into words, which is usually framed within the context of moving from one language to another. This is closer to an act of *transliteration*: this linguistic device gives textual form to the real fissures provoked by loss and dismemberment, which become translatable only insofar as what is untranslatable in them remains so. This literary technique brings to light the untranslatability of fractured bodies. It is no longer possible to simply translate things as they are; a transliteration of loss is needed, one whose resounding dissonance will take on a jarring linguistic form. Discovering the grammar of fractured bodies and disappearance is the goal of this literature of refusal.

The colonizer took many measures to give shape to an entire population of "*indigènes*": he dismantled the symbolic structures they all shared, erased their past, and undermined the languages they spoke. This gradually weakened the authority of its main taboos: incest and murder. This can be seen in the character of

Nedjma, who is born of a crime and unwittingly practices incest with her husband (who, as her father's son, is her half-brother). In this way, Kateb Yacine thrusts the reader into another world: one of lawlessness and self-enclosed communities where taboos have fallen. In Yacine's hands, Nedjma, the celestial incarnation of loss and heir of a whole series of massacres, has also become, like Oedipus in Sophocles' tragedy, the unknowing prisoner-agent of an incestuous relationship with her husband and her lover (both half-brothers). Yacine cleverly presents her to readers as all but mute while raising a constant suspicion around her father. Nedjma is a woman banished from speech; she can say nothing on the matter. Besides, only once does she express herself in the text. She is what remains after the colonizer's dogged dismantling of underlying tribal bonds. In traditional societies, the tribe is indeed the site of kinship and affiliation, binding together different generations and sexes, the living and the dead.[33] As one of the novel's characters says: "We are only decimated tribes."[34]

In both pronounced and subtle ways, *Nedjma* traces the breakdown of tribal society and shows its consequences on subjectivities and on the community. The disappearance of the subject and its referents creates, as Nabile Farès explains, a "foggy memory."[35] And it becomes impossible to approach this absence without disappearing in turn. "I know no one," Yacine writes, "who has approached her [Nedjma] without losing her."[36] The subject is bound up with passion, fascination, devastation for whoever lets itself be swept up by the beauty of the "ogress of obscure blood."[37] Here is how Nedjma is described:

> A woman perpetually in flight, beyond the paralysis of an already perverse Nedjma, already imbued with my strength, cloudy like a spring I have had to vomit in after having drunk from it; Nedjma is the tangible form of the mistress who waits for me, the thorn, the flesh, the seed, but not the soul, not the living unity where I could blend myself without fear of dissolution.[38]

In the star-studded text of *Nedjma*, incest is a direct result of the wreckage of genealogies wrought by the colonial order. Incest and murder spring from confusion, which hollows out the question of identity: once the name that establishes one's identity, recognizability, and affiliations disappears, one is left asking, "Who is who?" Yacine rebuilds a mythological world atop the ruins left by the War of Conquest, which he also sees as a way out of tragedy. Yacine turns obscene crimes into an esthetic of destruction. He suggests that the terror willingly imposed by colonialism presents a way of escaping tragedy. In this way, by reimagining the tragic, he restores the human at the center of history. He does this by mythologizing the "decimated" tribe in order to transform the real disaster caused by the French conquest: it is a matter of enveloping the disaster in tragedy and of fighting against the effects of terror, which lies beyond the tragic dimension.

One of the characters captivated by Nedjma asks: "[I]ncest is problematical ... what court?"[39] Indeed, what justice is there for destroying genealogical ties with a bureaucratic legal procedure? The "*indigène*" subject is excluded and stripped of its status as a human the moment it is excluded from the laws underlying the foundation of human life and generational belonging.

Nedjma cast a bright light on the scene of a dark drama dictated by colonial law. Doubt and suspicion reorganize the relations between genealogy, love, and war. In this way, the story lies out of reach, offering the reader no more than glimpses of what remains unseen. Incest reignites a ban on thinking. The story is one of a radical break. As things unravel with the ancestor of the tribe, the father becomes a "fake father."[40]

Incest evokes murder for Yacine. This is because, in the novel, Nedjma is also a place, a specific villa where a whole family committed suicide. On top of the ravages of colonization, there is also the death of the places to which a symbolic order is tethered. The law can only be imagined and caricatured after this point.

Yacine's text is therefore instrumental for understanding how the destruction of a symbolic order has greatly affected several generations of colonized peoples. Stripped of names, the subjects who go through this are also dispossessed of their own death. They wander through a zone of infinite time and space. Writing is capable of casting this lived experience – felt on the level of the real – as tragic: "That would-be suicide who comes to his senses no longer knows the illusion of dying," Yacine writes.[41] *Nedjma* is therefore a poetic translation of what took place at the time of the colonial conquest: the wholesale upending of the tribal system, which undergirded Algerian society at the time. Saint-Arnaud speaks to this in his description of a colonial policy designed to "decimate" tribes by smoking them out, eliminating those in power, appropriating their land, or spurring on fratricide – that tool of self-elimination that eradicates the "*indigènes*" while leaving the colonizer free of blame.[42] It is perhaps no coincidence that Yacine uses the same term as Saint-Arnaud to describe this: "decimate." This shows how Yacine draws on the signifiers of colonial destruction to construct a tragic story that recounts the murder and forced disappearance of the "*indigènes*."

The character of Nedjma is decidedly the product of a tribal structure that has collapsed under colonial rule. From a tradition of organized marriages between cousins within the same tribe comes the risk of bringing together brothers and sisters: if no one knows his or her father, then each person is unmoored with a false name. Incest is a permanent risk, as it is for Nedjma, whose body is both fully erogenous and entirely absent. She is struck with silence in the story but no one is silent about her. She becomes the origin of becoming, Borges' aleph, the traceless trace or *archi-trace*, always on the point of vanishing into nothing, as when she is abducted by the spirits of her ancestors. Luckily the star's white glow leaves a trace and so retains, like memory, this episode from history:

> We are only decimated tribes ... our ruined tribe refuses to change color; we have always married each other; incest is our bond, our principle of cohesion since the first ancestor's exile; the same blood irresistibly bears us to the delta of the passional [*sic*] stream, near the siren who drowns all her suitors rather than choose among the sons of her tribe.[43]

Disrupting genealogical ties: the effect of "renaming" Algerians in the 1880s

The logic of incest under colonial rule becomes clearer once one understands the role played by a new law established in March 1882 concerning the civil registry, which coincided with the "*code de l'indigénat*" put in place at the same time. These "reforms" were enacted a decade after the horrific massacres carried out by the French army of the recently formed Third Republic to suppress the uprising of Algerian "*indigènes*" led by El Mokrani in 1871. The colonial administration decided then to change the traditional practice of assigning each person a tribal name, which was deemed too burdensome for identifying individuals. The traditional practice relied on a sequence of names, followed by a reference:

name of one's father, grandfather, place name, and so on. Names were thus intimately tied to place, connecting by way of the father whole generations to the land and to history. With this ancestral practice, the name of the father (which is itself the name of his father, and his father's father, etc.), just like the land, is collectively owned. This was hard for the colonial administration to read. The autochthonous inhabitant knew how to navigate this system but the colonizer clearly didn't. Hence the effort to impose a French naming system, restricted to a first and last name provided by the administration, which didn't neglect filial ties altogether, but often severed ties with history and genealogy.

In line with other repressive measures, this is a forceful way of controlling the population by erasing any references made to the tribe and, by extension, to the father: In other words, this new French naming practice entirely abolishes the system that binds together the living and the dead. For Freud, "a man's name is one of the main constituents of his person and perhaps a part of his psyche."[44] The destruction of the name signals the death of a whole symbolic order wiped out by colonial law. And so individuals were renamed, or, rather, unnamed, on a massive scale by the colonial administration and therefore cut off from their respective genealogies. The risk, then, was that descendants from the same family may be given different patronyms, effectively making them strangers at birth and therefore subject to incest through marriage. This practice of assigning names with no bearing on genealogy or ancestral history deployed at times various strategies for containing individuals and eliminating tribal belonging, such as assigning everyone from the same village a name that started with the same alphabetic letter: names from the first village all started with A, those from the next with B, and so on, all the way down the line.[45]

Let's not forget the filthy and humiliating names that were imposed on them: Khra (shit), Boutrima (bearer of a small behind), Khris (coward), Khamedj (rotten), Resekelb (dog-face), Bahloul (idiot), Zani (fornicator), not to mention animal and other obscene names. These humiliating attributions were passed down for generations until there remained no traces whatsoever of filial genealogy. These were names that embodied humiliation, offense, and, above all, shame.

It was also common to provide a patronym based on one's trade or physical features. A passage in *Nedjma* shows how the children of banished fathers were required to bear the names of their trade, an effort by the French administration to co-opt them. They became judges of Arab tribunals (*kadi*) or *caïds* (chiefs) who controlled the "*indigènes*" for the benefit of the colonial administration – even today, the names "Kadi," "Bencaid," "Caïd," and so on, are still common in Algeria. "The ... condemned men's sons," writes Yacine,

> were still in their cradles when they were appointed acting caïds and cadis, thereby receiving patronymics corresponding to their future professions; hence the worst calculations triumphed even in the reparations that were made, for the name of Keblout was forever proscribed and remained a lamentable secret in the tribe, a rallying point for bad times. ... [T]he tribe's ruin was completed in the civil registers, the four registers in which survivors were counted and divided; the new authorities completed their work of destruction by separating the Sons of Keblout into four branches, "for administrative convenience."[46]

It took no more than thirteen years to establish this new civil registry. One can only imagine the shock and horror this provoked when considering that the previous system of filiation had been in place for thousands of years in Algeria.[47] The seriousness of this change cannot be overstated, severing as it did all ties to history and genealogy. The symbolic order that made family ties visible, and, by extension, enforced a taboo on incest, had been destroyed. The name given by the colonial administration marks the ruin of the living and the dead (the ancestral). Under the threat of serious penalties, one was forced by law to use one's new given name. This corresponded to a larger shift whereby the subject was becoming an object to be identified, catalogued, tracked.

Sons divided by their patronyms lost their genealogical ties. But if they no longer knew, over the course of a few generations, "who was who," the French administration could identify them, and therefore exercise further control over them by calling into question matters of inheritance and property. For the colonial administration, this law was also designed to allow them to identify assets, especially land collectively owned according to tradition: cutting individuals off from their communal history with fictive names facilitated real estate transactions and the expropriation of land, a goal underlying the very origin of colonization.

In his *Report on Algeria* (1847), Alexis de Tocqueville bears witness to the immensity of the destruction caused by the colonial administration's expropriation of land. From an economic standpoint, the sole purpose of colonial law was to upend the property rights that had been established in Algeria: falsifying land deeds, expropriating land, dismissing collective ownership, and so on. "The indigenous towns,'" Tocqueville recounts,

> were invaded, turned upside down, and sacked by our administration even more than by our arms. A great number of individual properties were, in time of total peace, ravaged, disfigured, destroyed. Numerous property titles that we had taken in order to verify them were never returned. In the vicinity of Algiers itself, the very fertile areas were torn from the hands of the Arabs and given to Europeans who, not being able or not wanting to cultivate them themselves, rented them to these same indigenous people, who thus became the mere farmers of the domains that had belonged to their fathers.[48]

The colonial administration's practice of unnaming, which began in 1882, was therefore an extension of its aim to revoke titles of collective property ("collective" in this sense corresponded to "familial"). Land lost the name of its owners, clearing the way for a sweeping appropriation of land that left no traces. The question of fathers, pointedly raised by Kateb Yacine in Nedjma's tragedy, is central to understanding colonialism's destruction of private and public spaces.

Subjective catastrophes and the disappearance of the father as symbolic reference

Nedjma, the "evil star of our clan,"[49] is the glimmering memory of what remains of fraternity when the father – the active link in a patriarchal system – becomes unknown. In this case, he can no longer recognize and name his children, nor can he be named as a father and find his distinguished place amid past and future generations. In traditional societies, the father mediates between past and future generations, between the world below and the world above, between the visible and the invisible: he plays the central role of intermediate link and negotiator between written and spoken law, defining the subject as the daughter or son of ... By dismantling filial ties, the civil registry occasioned the disappearance of the paternal function in society – while fathers were also actually disappearing.

In Kateb Yacine's novel, all the fathers of Nedjma's suitors have disappeared. The only father who remains is shared between her and her suitors (her half-brothers). And so the plot thickens. The reader is left to piece together familial relations, asking at all times, "Who is who?"

Yacine takes the disappearance of the father a step further by making the only living one a "fake father."[50] All the others can no longer be recognized or have disappeared forever. Children are haunted by the specter of destruction. Nedjma is an ersatz father, the trace of a missing person. She makes the blank space of *Memory's Absence* (Nabile Farès) shimmer against a black night. The enigma posed by fatherless sons is thus held in the sky's memory: "Where did the fathers go? Where in the world did they vanish. ... I think they took out the fathers and that the sons took out the sons."[51]

The father, excluded from the ancestral name and divested of any legal authority, is reduced to nothing. Disappeared, he no longer serves as an agent of the symbolic world or as the negotiator between two worlds (heaven and earth). As a result, the division between the visible and the invisible, material and immaterial, real and symbolic no longer holds. In his absence, the father is condemned to be but a shadow cast over his "bastard" sons (Jean El Mouhoub Amrouche).

Listen to the musicality of this beautiful passage from *Nedjma*:

> Do you understand? Men like your father and mine. ... Men whose blood overflows and threatens to wash us back into their old lives like disabled boats floating over the place where they capsized, unable to sink with their occupants: we have ancestors' spirits in us, they substitute their *eternal dramas* for our childish expectations, our *orphan patience bound to their paling shadow, that shadow impossible to dissolve or uproot* – the shadow of the fathers, the judges, the guides whose tracks we follow, forgetting our way, never knowing where they are.[52]

Yacine lets the fate of these vanished fathers emerge by way of suggestion. The reader is eager to know where these fathers went, but this information is withheld. These abandoned sons are left to grapple with the shadow of their fathers – disappearance appears at the end of this passage in the words "never knowing where they are."

Mohammed Dib inhabits the same fictional world as Kateb Yacine and is also plagued by the fate of sons whose fathers have disappeared. He examines the catastrophic consequences of this in his writing. According to Dib, sons are forever caught in the ravage wrought by omnipotent mothers. He writes about sons being unmoored from a paternal function: "The quest for a father fuels their worries and fantasies today – this father they never had to kill, as various colonial regimes from the past to the present already took care of this, effectively turning sons everywhere into orphans or into bastard children by confiscating from them their paternal figure. Colonizers made Algerians the sons of nobody."[53]

In line with this logic of disappearance and the orphaning of children, colonialism also led to the erasure of place names as countless locations were given French names. Thus, the new naming practice – the act of unnaming (abolishing local historical references) – targeted both bodies and land. The historian Daho Djerbal has argued that "the new official name lost all ties to past generations, to genealogical history, to the sacred; it was no longer a matter of history, or bound to sacred and mythological origins. It was a product of the colonial state in all its brutality." With respect to land, he adds:

> Thus, one finds on the geographical map of colonial Algeria toponyms evoking French victories, such as Aboukir, which refers to Napoleon's victory in Egypt on July 25, 1799; Arcole, his victory on November 17, 1796, as well as many others. Countless names of cities and villages, public squares and streets are there to remind Algerians of the glory of the victors of the War of Conquest, but also of other wars waged by imperial or republican France.[54]

The colonial practice of unnaming bodies and land could have no other effect but this. It imposed itself on the bodies and places of the autochthonous population while stripping the land of its historic and linguistic past. Dispossession is thus complete, triggering a widespread phenomenon of depersonalization. The relations between the sacred and the profane, the living and the dead, the visible and the invisible, material and immaterial were all upended. This shows the scale of colonialism's destruction of the symbolic order. It didn't so much repress history as *preclude* it, which is to say erase it irremediably. In clinical terms, this translates to the active *blanking out* of language, names, and history.

Writing against anonymous filiation

The change of patronyms under colonization ushered in a complete break with the tribe and the role it played as anchor and symbolic reference for living bodies. Mouloud Feraoun (1913–62), prolific writer of fiction who came from a poor Kabylie family, bears a patronym arbitrarily changed by the colonial administration from Aït Chabane (the name of his tribe). His writing illustrates colonialism's use of deception, which he experienced through his inability to escape the imaginary world to which colonialism had confined him.[55] Feraoun's writing seeks to negotiate between two fundamentally irreconcilable worlds. His work exemplifies the challenge of conveying loss and disappearance in writing. A narrative is deemed necessary for translating loss, but as Feraoun drifts into the realm of the imaginary, he also by necessity distances himself from experience lived on the level of the real. He believes being part of "the colonized" can be translated in a variety of ways. But this belief is a trap that further aggravates his pain. And for good reason. How can you create through writing an escape from History and narrative? And what languages are up for the task of describing dismembered bodies?

In his *Le Fils du pauvre*, which he began in 1939, Feraoun speaks about his impoverished childhood and the dissolution of his extended family.[56] He writes from the perspective of someone who has "assimilated." His education led him to become a teacher and allowed him to imagine "integrating" traditional practices within the Republic, a potential step toward social harmony. The French school system allowed him to escape poverty. His mastery of French, Kabylie, and Arabic made him believe in a collaborative future. But signs of this were nowhere present around him. The exterior world showed him how hollow his relationship to French and knowledge really was. Language does nothing to bridge the internal and external worlds, and this failure results in a linguistic self that is fractured. Feraoun found himself imprisoned within an "assimilated" language that only exacerbated the divisions he felt within. He had believed that the French language, providing symbolic relief, would make up for the language he had lost. But he discovered he was no better off with this language, and, worse still, it was a poor narrative vehicle for communicating the experience of "the colonized." In 1955, when he had given up on seeking common ground via the French language, he wrote:

> When I say I am French, I give myself a label that all French people refuse me, I express myself in French, I was educated in a French school system, I know as much as the average French person. But who am I, really? With all the labels that exist, how do I not have one? Can no one tell me who I am? Sure, I get it, I should just pretend to have one since everyone buys into that fiction. No, that isn't enough for me.[57]

Mouloud Feraoun was assassinated in 1962 by the *Organisation Armée Secrète* (OAS) along with five of his fellow teachers, some of whom were French.[58] In spite of his literary fame, the French language he had inherited didn't allow him to avoid being treated as "*indigène*." In other words, the French language didn't make up for the loss of his name and languages, which, for their part, were deemed deficient in "civilized" knowledge. The respective languages of the "*indigènes*" were effectively expropriated, and no other language could fill the void of, much less translate, that loss.

On another level, "selective" education provided in schools during colonialism helps explain how an inner divide first became internalized. At the time of Independence, only one out of nine Algerians were literate,[59] hence the profound gap that would come to divide the people and the new ruling class, which, as the *de facto* replacement of the colonizers, hardly offered any hope for reconciliation with the lower classes. Starting in the interwar period, the children of prominent "naturalized" figures were accepted in institutions where there was a mix of "Europeans", autochthonous, naturalized Jews, and autochthonous Muslims called "*indigènes*," "French Muslims," or "French subjects." It was more common for children to be in classes for "*indigènes*" in the big cities than in the country's rural interior. Their education was perfectly aligned with what was taught in the metropole, with a focus on the history and geography of France. The history of Algeria before colonization was ignored, reduced to a blank space. Young educated Algerians became unmistakably "Frenchified." Some young boys also had the opportunity to study the Qur'an in a *madrasa*, from 6 to 8 in the morning, before heading off to school. The way education was organized clearly illustrates the widening breach between two worlds: the familiar one handed down from previous generations and the foreign one imposed from the outside. The process of "integration" by way of "assimilation," to use the terms of the time, was an instrument of self-expulsion. This break from the familiar world was politically orchestrated.

Coloniality thus created a form of exclusion that worked on two levels: internal (within each individual) and external (us/them). The rift became so familiar that one tended to forget it was a political ploy. Mouloud Feraoun, for his part, refused this strict division between the colonizers and the colonized, which levels out differences and subjectivities in favor of a conveniently divided but unrecognizable mass. His *Journal* (1955–62) shows how this false division was but another instrument of coloniality. He was joined in this respect by the writer Albert Memmi, who, in his *The Colonizer and the Colonized*,[60] argued that the colonized and the colonizer are merely convenient categories that do not correspond to objective realities. They are products of a complex system (colonialism) which depends on a binary and exclusionary logic designed to divide people.

Hence the erasures and blank space that colonialism depended on to mask its own functioning. This effort to erase all aberrant elements shows not only that colonialism is guided by a totalitarian agenda, but also that this agenda originates from within the French Republic. Although the metropole turns a blind eye to these ties between the Republic and colonialism, Francophone Algerian literature puts them on vivid display. The entirety of the population ("*indigènes*," "naturalized Jews", "Europeans") was thus held hostage by a system that concealed its real name – even if, of course, the treatment of members of different communities remained unequal, as some kept their languages and names while others, kicked off of their own land and haunted by the reality of disappearance, inhabited a scorned tongue and lived with a fake name.

Isolating communities from one another is the visible face of a vicious and widespread practice of exclusion, affecting the very flesh and spirit of an entire people. It stands to reason that offense, experienced by all "*indigènes*" on the level of the real, also affected the wider population. This feeling of offense can be heard among those called, after the War of Liberation, "*pieds-noirs*" or, before the war, "Europeans." Did they trade their "European-ness" for a body part ("*pieds*" or *feet*) the color of Africa ("*noir*" or *black*)?

Jean El Mouhoub Amrouche: a broken voice

Learning French, cast as a tool of "integration" by turning away from the familiar, had the unintended effect of alerting young "*indigènes*" students to the very real threats posed to their own symbolic system. This awareness is all the more surprising given that French was meant to serve as a complete substitute for their respective languages, which would vanish without a trace. Just the opposite happened: these fading languages inflected French. Paradoxically, then, French absorbed and safeguarded the very mother tongues whose independent existence as a system of signs was being threatened.

From that point on, Algerian writers from the 1940s and 1950s strove to make this language one of recognition and of separation, a language that would serve as a burial ground for the dismembered and dispersed parts of stricken bodies, inviting restitution. In this way, these writers upturned the political organization of languages. Their use of French was a repudiation of the political order in France. Henceforth, identity became disassociated from language as writers found ways to make the unwritten blank space visible, a blank space clearly associated with spoken languages. Jean El Mouhoub Amrouche's literary oeuvre provides a vivid illustration of this.

Amrouche was sensitive to the piercing pain expressed in the Kabylie songs of his mother, Fadhma Aït Mansour Amrouche.[61] These songs addressed to no one in particular are marked by absence, an absence that resounds in his mother's hollow voice. This absence has shaped Amrouche's poetics, which can be described as a writing of disappearance. In other words, he writes to make disappearance present by welcoming these orphaned songs. His family fled to Tunisia, where they experienced exile twice over: first, from leaving behind their home in Kabylie, and, second, from living far from members of their Muslim community, exacerbated by the conversion to Christianity during colonialism.

Amrouche's style strives to turn this broken, "almost blank" voice, as he has called it, into a metaphor, one that will awaken in the reader a sensitivity to sound. He does this by transforming the experience of disappearance into linguistic pleasure. His writing must be read aloud to experience the musicality of his prose, which gives expression to the breaks in his mother's voice. It operates a shift from the visibility of writing toward the invisibility of the voice. The reading gaze suddenly transforms into a listening ear, reawakening the body's speech drive that once governed traditional societies for whom the word was law. The genius of Amrouche was to make French the acoustic vehicle for this sentiment:

> But before I could distinguish in these songs the voice of living people and those of wandering shadows, the voice of a land and a sky, they represented for me a singular form of expression, the private language of my mother. I have no words to express the thundering power of her voice, the virtue of its embodiment. She isn't even aware of this herself and for her these songs are not works of art, but rather spiritual instruments she used for particular purposes, like a wool loom, a mortar, a wheat mill, or a crib. Hers is *almost a blank voice, uninflected, infinitely brittle, on the verge of breaking, a little shaky and each day growing quieter, losing its stability with each passing year.* Never any explosive moments, not the slightest concern for external experience. With her, everything is softened, interiorized. *She barely sings for herself; she sings above all to soothe and endlessly revive a pain sweetened by its very incurability, a pain intimately bound to the rhythm of her sipping of death with each breath.* It's your mother's voice, you might say, it's only natural that you're obsessed with it and that it awakes in you the muffled echoes of your childhood, or the interminable weeks when we regularly confronted absence, exile, or death. You're right. But there's something else to it, too: floating on the surface of this voice awash in a blur of languages is an infinitely retreating nostalgia, *a nocturnal light from beyond which emerges the feeling of a presence at once close and unreachable, the presence of an inner land whose beauty is revealed only insofar as the land is lost.*[62]

The mother "barely sings for herself": did colonization leave the colonized with no one to turn to? Indeed, it wasn't so much that languages were lost, as families continued to live, think, and feel in these languages. Rather, what makes each language the site of knowledge and interpersonal exchange was "decimated." This aspect cannot be recovered. All one can do is find a style of writing whose musicality can make a splitting and haunting pain resonate, bearing witness to what has disappeared.

Amrouche turns the disappearance marking his mother's voice into a metaphor of loss harbored by another tongue. There is a need for writing to tend to the damage done to the symbolic order.

Confronting an army, the poet uses language and culture as his or her weapons in the "civilization" process.[63] Not that this language, which is also a vehicle for savagery, is embraced without ambivalence. But French, the language of the Other, offers a circuitous route back to one's mother tongue (the private Other) insofar as one can use it to bridge division. It doesn't rely on translation, but rather on transliteration, on transmitting lived experience – felt on the level of the real – where other languages fall short in fulfilling their civic duty. What matters most for these writers is to recreate a living language, to provide a tomb for the missing (bodies, names, land), to give rise to a sort of indelible textual memory. The "*indigène*" inflection of French can be perceived in a text's margins where other languages, sensory experiences, and songs thrive. In an interview in 1963, Kateb Yacine put it this way: "We were invaded by France for thirty years! What did we get out of it? Ruins, corpses! … But also a language, a culture. If we were to throw that away, we'd really be behaving like savages!"[64]

Francophone Algerian writers, these scribes of disappearance, created their own hybrid poetic language to resist erasure from the Republic's "national" language. In this way, they allow us to rediscover the real occasioned by mother tongues buried in shame and disavowal. In the oppressive context of colonialism, this encounter with French made other languages and their symbolic body all the more distinct. There was no longer a clear division between the familiar and the foreign. These writers thus developed a linguistic approach that exposed the tension inherent in all languages, a tension between opening (via poetry) and closure (history imposed by others). Using language to free oneself from colonial subjugation while freeing language from its "national" captivity – such is the contribution of Algerian literature, which, presently, it performs in two languages.

Approaching French from this angle, Kateb Yacine speaks of his enthusiasm for the French Revolution. He explains how learning about this history of France forced him to redefine his relation to his own history, in spite of the fact that, as a student, he was compelled to accept French history as unquestionably his own. Unexpectedly, altering the master's language seemed to offer an alternative to "assimilation." Hammered with French lessons, children came to discover how, when viewed in a colonial context, the Republic was filled with glaring contradictions:

At French school, I learned something fundamental that I never got from Qur'anic school. It was a sort of paradox, if you will, but one that marked me. I had an insatiable appetite for learning, I devoured books, and think I was quite precocious; I was reading Baudelaire. ... Then the French Revolution, which really spoke to me, became my obsession.[65]

3
Colonialism Consumed by War

The French of Algeria, especially those
with great wealth, have convinced themselves
that 1789 never happened and that its principles
don't apply to Algeria.

Ferhat Abbas, 1980[1]

The truth is right there with its droopy, sleepy eyes. It is
looking at you, unshakeable.
Not even its ugliness affects it, its own ugliness.

Malek Haddad, 1989[2]

If the colony represents a scandal for the French Republic, it is above all, as we've seen, that of a *rogue child* born of a strange love between the Republic's democratic principles and its capitalist discourse (willfully exploiting and commodifying human beings). This rogue child of the Enlightenment suffers from a non-separation. It rages deep in the bowels of the Republic. As the practice of disappearance reached its peak, the war inevitably escalated. This explains how the massacres of May and June 1945 in the northern region of Constantinois were a decisive turning point, prompting nationalist militants to put an end to colonization. The horror of war was all the more pronounced as the colony and the Republic suffered to stand as separate bodies. Instead, they merged into an indistinct, all-consuming body. Erasure sought to make boundaries, borders, and differences all disappear. Diversity is a stubborn problem for coloniality, which uses every means at its disposal to abolish it.

1945–1954: the necessity of war

From an Algerian perspective, this separation is experienced as a reintegration of the beaten and mutilated body. For the French, however, they were now the ones dealing with the feeling of offense and dispossession – the separation was thus felt as an amputation as the so-called "Europeans" fled Algeria in 1962. This impossible separation explains how Algeria remains the site and name of a disaster affecting both countries to this day.

The divisions within the colonial system, as we've seen, worked to restrain (if not outright abolish) diversity in the discourse and practice of subjects. The goal was to create a uniform, though paradoxical, entity: at once fundamentally indivisible and yet starkly divided between "us" and "them." In reality, this colonial construct was nothing more than a political fantasy, since it was belied not only by the plurality of its "dominant" class (French, Corsican, Italian, Spanish, Maltese, Swiss, British, German, etc.), but also by their positions on colonialism and their relations to the autochthonous populations (Muslim and Jewish) and to the land. In other words, the division between both parties was created as an instrument of power and domination designed to erase internal differences within each group. Complete uniformity was the goal of the colonial order. Division was viewed as a threat to an undivided state that depended on the absence of conflict between both parties within the colonial system. After several generations of colonization, the deception was complete, with no one noticing, much less questioning, the single body of colonialism.

Decades of murder, repression, humiliation, contempt, and offense nevertheless led to an outbreak of violence for both parties. Of course, this violence was markedly asymmetrical, obeying a different set of protocols and strategies of war, not to mention a different logic of accounting for the living and dead. The violence against bodies both real and symbolic through mutilation and dismemberment led to a bloody collapse of colonial society. Before the war, the division between "settlers" and the "*indigènes*" was maintained by the colonial system, but after World War II and its aftermath in Algeria – the massacres of May and June 1945 – cracks started to show in this façade and a major shift occurred. The division between "us" and "them" was no longer tenable, and nothing could change that. The rattling of these divisions, which soon became irreconcilable, triggered the War of Liberation.

The War of Liberation, from November 1, 1954, to July 5, 1962, signaled that Algerians were no longer willing to accept the practice of dispossession (affecting language, history, and religion). Their revolt, which led to the Revolution, was the logical and irremediable consequence of a widespread feeling of offense. Murders, arrests, torture, hunger, constant humiliation – none of these was enough to amount to a demand for liberation. The novelist and psychiatrist Yamina Mechakra expressed it like this in her novel *La Grotte éclatée* (1979): "Freedom? I swear to you I love it more than I am able to. I had a hard time finding the right words, paralyzed as I was by hunger and emotion."[3]

Firsthand accounts from those who fought in the war display an unprecedented level of passion and emotion. They had embarked on an irreversible path of *political action*. Decades of fear and terror shouldered for generations were shaken off in order to write a new page of History. Young nationalist fighters were fueled by a deep-seated rage and anger. They had reached a point of no return, leaving armed warfare as the only solution.

The political aim of this war, called on the Algerian side a "War of Liberation" or "Revolution" (*tawra*), was to re-establish what was destroyed under colonialism. This was a matter of restoring subjectivities, language, and tradition; in short, a return to the "self" at the heart of a burgeoning Algerian Republic. On the French side, this war remained nameless until it was granted official recognition by President Jacques Chirac in 1999. During the war years, it was merely considered an outburst of violence stemming from both sides. One of Kateb Yacine's characters, Mourad, speaks tellingly of Suzy, the daughter of a colonizer who is also his boss: "[I]t is as if we should never be in the same world together, except because of violence or rape."[4]

At the same time, colonial divisions infected various nationalist movements as well as the French political arena and its treatment of the war. The *Front de Libération Nationale* (FLN), which launched the war through a series of coordinated attacks, was forever plagued by its attempt to squash political, historical, and linguistic plurality within its own ranks. The Revolution, seen as breaking free from colonialism, remained nevertheless divided between a pluralist, secular nationalism and a nationalism set on purging differences, a purified one belonging to the *Umma*, or community of Muslim believers.[5] This political fracture over Algeria's future served to mask deeper internal struggles. At the heart of these struggles was a search for origins, in particular a single origin that could be reclaimed by all in response to the "genealogical wound" from which many suffered.[6] Fratricide, advanced, as we've seen, as a war strategy by French soldiers in the early days of colonialism, became a way of asserting power internally (see chapter 5 below). Thus, another war was being waged behind the scenes during the War of Liberation, a fratricidal war fought internally and among families.

Since the uprising of November 1, 1954, which marked the beginning of Algeria's external war, there has been an endless supply of stories and accounts of this war's history such that the other war, the Internal War fought over origins and legitimacy, has remained in its shadow. But both wars are inextricably bound together, and should be studied in their interrelatedness. The history of Algerian "nationalism" thus conceals within it a "cryptomnesic" memory, which is to say a memory that is hidden and encrypted (see the following chapter). Indeed, the children born under colonialism inherited this struggle to develop a political thinking that embraces difference. The feeling of offense and hatred for the republican pact at the origins of colonialism stood as a major obstacle to fraternity. How could militants and combatants develop this kind of political thought when men and women had been banned from their land for generations? War became a necessity to restore a humanity broken by decades of violence and the near constant practice of disappearance (of both people and material, spiritual and mytho-symbolic goods).

The practice of disappearance targeting the "*indigènes*" is clearly expressed in the words of Lieutenant Colonel Lucien de Montagnac (1803–45): "Any population who refuses our conditions must be eradicated," he wrote on March 15, 1843.

> No one, regardless of their age or sex, should be spared. Everything should be raided, seized. Grass shouldn't even grow where the French army has stepped foot. No matter what our philanthropists say, the end always justifies the means. I have warned all the good men that I have the honor to command that if they bring me an Arab alive they can expect to be beaten with the flat end of my sabre. ...

> And this, my good friend, is how you defeat the Arabs. You kill all men fifteen years old and up, take all the women and children, take control of their dwellings, send them to the Marquesas islands or elsewhere; in a word, abolish everything that doesn't crawl before us like a dog.[7]

The War of Liberation was meant to put a stop to the practice of disappearance and counter its long-term impact: the wholesale whitewashing of the land and the mental landscape of its inhabitants, both by repopulating it with white people and by blanking out the land's history through erasure. Since France's conquest of Algeria, the practice of disappearance, carried out in many forms, was the cornerstone of colonial policy. This explains why young Algerians who were targeted or who witnessed the massacres of May and June 1945 were ready to give their lives and use their bodies as weapons to put a stop to the forced, century-long, decline of their people. The War of Liberation springs directly from the unspeakable violence committed during the unnameable War of Conquest. The young combatants who led the insurrection on November 1, 1954, had grown up haunted by the orchestration of disappearance during this long war. This generational memory is essential for understanding the historical context of the War of Liberation. Nationalist militants were driven by the need to rewrite History at the site of colonialism's erasures. A pointed response to Montagnac's 1843 letter can be found in Albert Camus' *The First Man*. Getting ready to leave Algeria at the end of the war, a settler says: "Since what we made here was a crime, it has to be wiped out."[8] This isn't a case of denying colonialism's impact; rather, he is deliberately disavowing it, a sort of "I know good and well, but still" (Octave Mannoni). This statement implies that the speaker acknowledges the crime *and*, at the same time, that he wishes to know nothing of it – unlike denial, which never admits to the crime in the first place.

The impossibility of forgetting and madness, a "remedy" for disappearance

For the colonized, it is bodies above all that bear the memory of widespread mutilation. The body's material presence on the level of the real becomes the site of memory. In her novel *La Grotte éclatée* (written in 1973; published in Algeria in 1979), Yamina Mechakra makes this colonial branding visible. She speaks of bodies ripped apart by combat and torture during the War of Liberation. *La Grotte éclatée* is not a story, but the diary of a body torn to pieces. The narrator, a nurse who is part of the resistance, sees and cares for these bodies, but repeatedly experiences firsthand the dismemberment and madness of the "carnage": "Every time I looked up, I saw carnage. Mutilated bodies everywhere. What crime did these men who were born to live commit? Why had they been killed? What wrong did they commit, these men who were not even given time to be loved?"[9]

The narrator lives in a cave with fighters who are seriously wounded, dying, or delirious. This is where they go for care. The description of all these dismembered bodies evokes another episode in this long history: that of the *enfumades* during the War of Conquest. The fact that these mutilated bodies are found in a cave recalls those earlier massacres. However, Mechakra never directly alludes to the *enfumades* that wiped out entire tribes. Bodies disappear, but History is preserved on site. Yacine wrote a preface for the novel and points out that these caves had also served as underground passages during the war against the Romans (see chapter 8 below). But in this case is memory bound to time, place, or History?

During the war, Algerian bodies – those of both men and women – were subjected to torture on a massive scale. These were people whose bodies were their only defense. But, as we've seen, the dismemberment and display of mutilated bodies dates back to the War of Conquest. Thus at the very outset, colonialism was enforced through a very particular treatment of bodies. The War of Liberation simply offered an excuse to institutionalize existing practices of torture. Mechakra depicts the horror of these practices:

> All of Algeria bled from its wound
> Villa Susini; naked girls are drowned in bathtubs
> Villa Gras; men are suffocated
> Electric shock torture
> A body drenched in gas burns
> Torch flames split open chests and arms
> Whole bodies are cut up
> Hands serve as anvils for axe blades
> People are strangled
> All of Algeria was screaming under torture.[10]

In *La Grotte éclatée*, bodies are dismembered, the dead are unburied, and children are burned with napalm by the French army. The consequences of these historical facts that took place in 1958 are spelled out in the novel: the experience of an internal disappearance is shown to plunge one into delirium: "I almost went mad and sought madness as a remedy."[11] These mothers who have no more tears to spare for their dead children (killed regardless of whether they were combatants or civilians) lead the reader toward another world, a world of savagery. Psychologically, the war has never let up for the colonized since the start of colonialism, it has continued to rage even during moments of respite. Mechakra indicates this when she states: "We no longer have time to reinvent ourselves."[12] The cave was already filled with the memory of the *enfumades* and its mass of convulsing bodies. The cave also symbolizes the collapse of the tribal bond.

The novel is dedicated to grieving those who disappeared, the living's unburied dead. Grief stares at us "with the eyes of memory."[13] The narrator buries the bodies she finds, but so many remain unburied that she is thrust into a space filled with specters. This is the mark of her madness: "Nobody," writes Mechakra, "will know that the young man, who left us at the dawn of his youth, died without a grave; this land that he defended turned its back on him."[14] Mechakra brings us close to the mothers who, unable to bury their sons, go mad. For the living, this experience of bodies torn to pieces, burned, cut up, is all that remains of the disappeared. Indeed, burial gives the dead body a wholeness, a memory, whereas disappearance brutalizes memory, leaving the living in a state of suspension between life and death. Disappearance eats away at the spirit of the living and turns the catastrophe into an eternal present: "Her memory was scarred by burn wounds."[15] Is this not the "dark memory of times past"?[16]

Still today, there is no list of the names of those who disappeared in Algeria. The disappeared have yet to die. The catastrophes wrought by the War of Conquest are rarely mentioned, and the fact that a third of the population disappeared seems to have been forgotten. All the attention is given to the War of Liberation, praising those who fought as heroes and lamenting the number of lives lost. How can we explain the odd "disappearance" of the War of Conquest while the War of Liberation refuses repression?

Colonialism and its treatment betray an inherent obsession with memory. In Algeria, disappearance has written itself into history to such an extent that even creating an archive is impossible. *La Grotte éclatée* offers a textual memory – one that is filled with its own blanks – for fragmented bodies, and serves as a prayer for the disappeared. Mechakra makes the history of a colonial madness *impossible to forget*. While the narrative and memory of the first war are forgotten, bodies remain marked by this catastrophe. Did what escaped memory find refuge in the body? In the colonial context, memory is in a paradoxical state of confusion, caught between erasure and the impossibility of forgetting. What to do with the memory of these bodies torn to pieces? How to facilitate their entry into a healthy oblivion? We will see how mutilated bodies and disappearance still belong to a practice of non-forgetting in Algeria today (see chapter 6 below).

The impossibility of forgetting plunges one into madness, which is itself a remedy, according to Mechakra. The psychic pressure causes one to enter into a spectral universe filled with hallucinations as the body's strangeness and quirks come to occupy the psyche and fight against the body's own annihilation (disappearance). The narrator knows that, to stay alive, forgetting is the sole remedy and yet, for her, nothing can be forgotten. Forgetting can't happen on command. For the narrator, forgetting would entail making what happened disappear. In other words, she'd re-enact the very crime she is trying to forget. Mechakra offers a way out: an encounter with another (in this case a healer) who acknowledges the disappearance, mutilation, and cries of the fragmented bodies invading the mental space of the living. From this encounter a "memory of the Absent"[17] is born, which offers a burial for the homeless bodies that have taken over the minds of the living:

> Sick, sick from head to toe. I went to see a healer. I told him I was tired of the blood, that I shit blood, vomited blood, that my head hurt and that I wanted to forget everything. ... And if I can't that he grant me with his powers the illusion to believe in myself. He put his hands on my head and for a few moments, under the weight of my pain, I was able to forget. This was good for me and I returned to see him every day ... every day. Each day he'd put his hands on my head and each day I would move away from a source of helplessness that was dragging me into the depths of misery. And then one day, my healer sang to me the songs of people who were dying and who were scared to die. He sang me songs from the living who were scared to go on living. I was neither afraid of dying nor of living. I let him sing because his hand on my head calmed my memories, quieted my

pain, because, with his hand on my head, I was less horrified by my own solitude.[18]

Silencing the unforgettable mutilation of bodies

The colonial occupation was thus designed to mutilate bodies and memory. Over time this gives rise to a troubled and confused relation to the past. Therein lies the principal division between the French and Algerian political orders, still felt half a century after Independence: on one side, an unambiguous disavowal of this history of crime and massacres, aiming to keep the population divided over its own history; on the other side, a wide-scale repression of the vast destruction caused by the War of Conquest while bodies remain marked by its effects and hero narratives about the War of Liberation abound, narratives that displace the memory of other fratricidal wars (such as that between the FLN [*Front de Libération Nationale*] and the MNA [*Mouvement National Algérien*] as well as an internal conflict within the FLN – see chapter 5 below).

Under colonialism, the sites of forgetting and those which were *impossible to forget* were clearly demarcated. The absence of memory for one translated into an excess of memory for the other, and vice versa. This raises the question of how the political order determines what gets remembered, and what is the relation between law and memory. Clearly, historical facts alone cannot explain what gets remembered, which is a process subject to detours, distractions, and blank spaces. For the mutilation of bodies is a practice of disappearance that triggers repression: it is a way of inscribing the dispossession of self onto the body and ensuring this gets passed down to subsequent generations – this is notably the case with rape. Colonialism's treatment of bodies is the best way to occasion a narrative rupture.

Still today there is very little work in Algeria on the psychological effects of torture and its impact on future generations. The men and women who were involved in these wars are rarely afforded the occasion to discuss this. An exception to this is the account of Louisette Ighilahriz, who, in her book *Algérienne* (2001), describes the torture she was subjected to during the War of Liberation and the effects of this on her body.[19] In an interview after her book appeared, she noted how what she discussed was still taboo in Algeria and that people tried to shame her for talking about it. Indeed, although it is readily acknowledged that torture was a common practice during the War of Liberation, no one is permitted to speak about its effects on subjects. Hero narratives continue to mask the real damage caused by war and colonialization, which remains cloaked in silence and secrecy: the Algerian political order strives to keep what took place – what remains unforgettable – silent. The French political order takes the opposite approach. It prefers to ignore entirely this part of history and the memories held of it by a not insignificant part of its population.

The French history of Algeria and the Algerian history of France demonstrate that the political order always determines to a large extent what gets remembered and what doesn't. Both countries are grappling with the legacy of colonialism, and in both cases politics inevitably shapes memory. In Algeria, it hasn't been easy to subjectivize this history, as History continues to be treated as a text written by the political order. In France, in spite of the work of countless historians, colonial history remains barred by the political order from entering into collective memory or public debate. It is often treated as a matter that only concerns "minorities." The blank space to which the monarchy (see chapter 2 above) was subjected continues to expand within the Republic. Is it not in that same blank space where colonialism resides nameless? If so, the trouble of inscribing memory (be it an excess or a lack) afflicting both Algeria and France isn't so much a deliberate refusal to acknowledge history. Rather, it seems more precise to speak of colonialism's disabling of the symbolic structures on which traces of memory get inscribed. This enabling and disabling of memory is part of a larger goal of colonialism to appropriate History and make it an instrument of politics. The lasting effects of this can be seen in both France and Algeria.

Murdering "*indigènes*" was part of the culture of the Republic. This legacy has clearly impacted the processes of historical-subjective repression. The mechanisms of memory were disabled in order to allow blank space to spread, which undermined one's ability to control memory and forgetting. Memory itself ends up being subjected to erasure and the disappearance of traces. What Frantz Fanon says of colonialism's impact can be applied to a wider field that encompasses all components of the colonial system – "colonizers" and the "colonized" – and hinges on the blurring of boundaries between memory and politics: "French colonialism," he wrote, "has settled itself in the very center of the Algerian individual and has undertaken a sustained work of cleanup, of expulsion of self, of rationally pursued mutiliation."[20] This expulsion underlies the political order's relation to memory and history in both France and Algeria.

The law of murder at the heart of a regime founded on exception in colonial Algeria wiped out the culture to such an extent that this legacy can only appear as blank space – a blank space expressed through a lack and excess of memory. For it isn't so much the War of Liberation that has remained unnamed but the murders and brutality carried out under the orders of the French Republic and its savage treatment of the bodies of the colonized for over a century. Over the course of several generations, the madness of this practice gave rise to a split memory: on one side, traces of memory allowing for the return of the repressed, and the rewriting that accompanies it; on the other, the eternally unremembered, with the possibility of repression, and its corollary, forgetting, foreclosed, prohibited. The ravages of memory's dismissal prevent the writing of the future from reinventing history.

In this blank zone of memory, history is immobilized: it becomes precisely that which cannot be changed or modified from the past. It grabs hold of bodies beyond discourse, speech, and circuits of recognition. In this dismissed zone of the psyche and of history, there is no return of traces, but rather a struggle to create a living trace that would follow its own path. The writing of history is dragged down by a hole that spits out in random gobbets a wandering mass of "encircled bodies."[21]

Toulouse, 2012: the return of murder

Fifty years and one day after the Évian Accords were signed in 1962, Mohamed Merah, of Algerian descent but a product of French society, heir of the blank spaces of Franco-Algerian history, opened fire in a Jewish school in Toulouse, killing four people, including three children. An appalling act that reveals an excess of memory in a place where, notably, the political order strives to keep memory in check. These murders, as well as those of three officers that he had committed a few days earlier, suggest that the reality of violence transcends time, condensing past, present, and future into a single disaster. Franco-Algerian history proves once again that it can only be experienced through war. Is this history condemned to a backward-facing future?

The murders committed by Merah seem to show that the work of memory has two sides. On the one hand, the date's overdetermined symbolism suggests this memory has trouble finding expression, which questions the whole notion of *fatum* (fate or fatality) surrounding the War of Liberation. On the other hand, it is as though these murders failed to recall the history between the two countries. In this respect, the debate over Merah's burial site – France or Algeria – is telling: France demanded he be buried in Algeria; Algeria, for its part, refused to accept the killer's body, arguing that he wasn't an Algerian citizen. Serving as a loud reminder of a bloody history, Mohamed Merah's body was thus viewed as an unwanted burden by both countries. His actions are borne by an excess of memory, which struggles to be translated. And these blistering memories dissolve into blank space.

It is a strange phenomenon where an excess of inscription leaves no traces. We are dealing here with a mechanism that disrupts the historicizing processes of memory: memory is still at play, but it is dislodged and disappears from its familiar dwellings in narrative and remembrance. Historical reference is struck by erasure. This gap, which causes a blank space of history and memory, makes it difficult to establish an archive of memory, a necessity for historicizing and subjectivizing. However, this blank space is grasped in speech's dismissal, made visible by Algerian writers. And it exists in this form in the current political discourse of the dominant French media and in numerous immigration policies in France.

Coming from a different angle, the comedian and director Daniel Mesguich (born in 1952 in Algiers) has made pointed remarks concerning this madness of memory. In 2016, he told Leïla Sebbar in a letter:

> You asked me to tell you, my dear Leïla, what the Algerian war was like for me as a child. ... But in order to recount something, this something needed to have occurred, don't you think? First of all, it would have to constitute a thing, a stable and consistent object to be exact, and then one would have to be able to grasp it, observe it, and then remember it. Alas, the Algeria war is nothing but a vast blankness for me. Or, rather, a sequence, even a symphony of blanks. All kinds of blanks, superimposed, arranged in an overlapping

> pattern, or threaded together. ... Generally speaking, it is absent from my memory. The first absence.[22]

Mesguich foregrounds, as Lacan would say, the traumatic absence [*troumatique*] of this war by its lack of memories, beginning with its inability to be named.

Thus, Mesguich suggests, for "Europeans," colonialism and its collapse represent a giant blank within the Republic. War is a bloody operation. It takes place between a call for separation and a search to recover a lost body. Algerians wanted to break free from colonialism and regain everything they had been stripped of, beginning with the symbolic order that had been taken from them (language, history, culture). In this way, liberation offered the promise of "reintegrating" what was lost and of escaping the blank space to which they had been confined for several generations. For representatives of the French Republic, it was above all a matter of refusing this deliverance, which was perceived as a serious threat toward the body republic. The collapse of the colonial order shook the very foundation of the Republic. And for good reason, since, as we've seen, colonialism occasions a break with the republican pact. The *rogue child* (blank space of the Republic) was returning to swallow up his progenitor right when it was to be separating from him in agony.

The acknowledgment of two distinct entities led to a profound feeling of deception on both sides. The "single body" proved to be a republican illusion that wasn't shared. The war hinges on this impossible separation, which affected both countries. Countless atrocities were committed on both sides as the fall of colonialism triggered an unprecedented bloodbath. As Benjamin Stora explains: "For the French it was a 'nameless war' and for the Algerians a 'faceless Revolution.'"[23] Put this way, Stora calls attention to the impossible separation between the name and the body; at play is a unified mass whose separation is experienced as a form of dislocation.

Reduced to the status of unburied bodies, Algerians weren't even given the courtesy of a death toll. This offers some perspective on the debate between the two countries over the number of deaths caused by the war. The 1.5 million estimated by Algerian authorities takes into account the nameless bodies that disappeared into mass graves since 1830. French historians, for their part, believe that around 350,000 Algerians died during the War of Liberation.

If, for the French, this war "lacks memory," as Daniel Mesguich puts it, for the Algerians, an endless eruption of blood and violence surges up from this absence.

The cult of the dead who are endlessly glorified hasn't allowed for a full reckoning of the absence provoked by massacres, murders, and destruction. This is an effort to patch over an absence with an overabundance of memory – a task diligently pursued by the political order in post-Independence Algeria. This shows how pleasure built up over time from destruction is released at the very site of loss, which is not unlike the cult of the sacrificial position. Nevertheless, this excess of memory contains its own blank space (see chapter 5 below). Seen as the underlying cause of each and every problem, the destruction caused by France's colonization of Algeria, though unquestionably significant, also came to serve as a distraction from other pressing questions of responsibility.

Constructing the "nation"

The War of Liberation casts itself as revolutionary in the fullest sense: positions were radically reversed, those belonging to the "*indigénat*" became citizens, languages that were once banned were re-embraced, and one had the right to practice one's "own" religion. However, in post-Independence Algeria, this subversive goal soon gave way to that of complete integration. The Revolution sought to restore a dispossessed, beaten, dismembered "us," an "us" who had failed to pass the purity test of integration.

Following a series of massacres and atrocities committed during the War of Liberation by the *Armée de Libération Nationale* (ALN), the FLN, the lone party at the head of post-Independence Algeria, quickly took measures to suppress and purge any dissenting elements from the party line that were perceived as threats to its unity. The women and men who promised to reanimate history by reviving languages and marginalized histories, religions, and nationalities were seen as a threat to be eliminated. That included Algerian intellectuals as well as pro-Independence "Europeans" and *harkis*.[24] Despite the differences between them, all of these groups represented a danger for the FLN's ideal of purity. An attack on alterity was unleashed as the reign of homogenization was established. The nationalist ideology saw itself as countering the colonial ideology point by point. But this opposition proved to be but an extension of the colonial legacy.

The Revolution's promise for change was thus hijacked [*détourné*] in order to pursue precisely what people claimed they wanted to move away from. Soon after Independence, *détournement* became a central instrument of power to suppress any signs of subversion and a forceful but silent ally of censorship. Colonialism's repressive and totalitarian system (targeting the "*indigènes*") was revived by the young Algerian Republic. Censorship continued to refine its orchestrated silence by expanding its control over speech and thought and maintaining unbridgeable divisions between the self and the other.

Glorifying those who disappeared and making disappearance a national cause created a counter-response: "Ancestors grew ferocious."[25] Disappearance became an instrument of state terror (see chapter 7 below). The "traumatic absence" ["*troumatique*"] occasioned by the colonial practice of disappearance continues to be rehearsed in Algerian politics today while ignoring its origins. The consequences of this have been disastrous, starting with the fact that no one counts the missing (see chapters 7 and 9 below). Kateb Yacine speaks of the impact of disappearance by insisting that the practice itself hasn't disappeared (quite the opposite, as it seems to have become a permanent feature of Algerian history, politics, and society):

> Thus glory and defeat have founded the eternity of ruins upon the growth of new cities, more alive yet severed from their history, deprived of the charm of childhood to the benefit of their ennobled specters, like pictures of dead brides that make their living replicas grow pale; what has perished flourishes to the detriment of all that is yet to be born.[26]

Yacine believed that theater offered a unique intertextual site where the individual, the larger public, and the political order could come together to confront the erasure of catastrophes. This stands in stark contrast to the art of *détournement*. And yet, despite his efforts as a writer and playwright, his work could not overcome unbridgeable divisions. Like many intellectuals after him (Francophone and Arabophone), Yacine's work was never widely embraced: "How can writers avoid talking over their own people, writers who must often pass through France first, who must travel the world and return like an echo to those who, among their own people, are open to him: which is to say, at the time being, a small minority."[27]

The absence of a space of reception for speech, discourse, and the work of culture is a tragedy that still afflicts Algeria today, one that serves to prevent a plurality of voices from emerging. The deep-seated feeling of forsakenness that results from a lack of reception is felt on all levels of society. It is played out within each individual subject, no matter that person's sex, language, or generation, regardless of whether he or she be a leader or an ordinary citizen, a caregiver or a patient, a writer or a reader, an Arab-speaker or a French-speaker, or even a professor or a student. However, writers are relentlessly pursuing the work of culture and seeking new opportunities for subversion. Writing remains an urgent task in the land of the LRP, one that seeks to establish a different relation with History, with memory, and, ultimately, with politics.

The writer's pressing need: transform disappearance into absence

Despite Independence, writers continue to write today against other forms of disappearance and subjugation, this time provoked by a "national" heroic narrative. Writing continues to be a means of resisting, of bearing witness, and of documenting. It is a matter of creating spaces of writing that suspend the otherwise inevitable fall into the gaps of memory. Literature stands in opposition to the political order by giving body to disappearance and by recovering it at all costs. In Chawki Amari's *Le Faiseur de trous* (2007), the following exchange takes place:

> "Do you know the joke about the truck that was carrying holes?" Moussa asks Aïssa, holding a bunch of cables.
> "No," Aïssa responds, well aware of the joke.
> "Well, one of the holes fell from the truck while it was moving."
> "So what happened?"
> "Well, when the driver saw that, he stopped and backed up."
> "And then he fell in. He fell in his hole."[28]

Filling this hole as a form of recovery depends on the most dangerous illusions. Political leaders set on building the "nation" did just this. In contrast, the writer seeks to get out of the blank space by filling the space of language with vacancy. To accomplish this, the writer makes loss a central component of his or her writing, thus preserving disappearance. Thus, loss and disappearance are in stark opposition if the subject allows loss to disappear without trying to recover it.

Algerian literature has allowed us to identify the consequences of colonialism and its impact on subjectivities and the political order. As we've seen, the colonial order sought to diminish – or even exterminate – fathers and the paternal function. These disappearances gave rise to countless fratricidal wars, including the Internal War of the 1990s. The sons of the Revolution were plagued by a terrible feeling of illegitimacy, which can be tied to the disappearance of their fathers and the destruction of family lines. This in turn disrupted the means of recognition within the symbolic order – the paternal function that names, recognizes, and allocates a place within a lineage. Does this disruption thrust one into a state of mourning that has never begun?

The enduring practice of disappearance in Algeria that finds its origin in the colonial period has all but suppressed the possibility of a plurality of historical traces from emerging. Alteration – the effect of encountering alterity – remains in particular a threat, as one is suspicious of a foreigner's intentions: he or she can become an agent of disappearance at any moment. Through successive periods of colonization (Arab, Turk, French), the encounter with alterity has over the course of the centuries undermined existing foundations rather than bolstered them. Never have different historical periods, or differences in ethnicity and religion, been welcomed and integrated. Rather, these have been systematically excluded in order to construct a homogeneous society, a society that neglects its own history and alliances: "This world isn't one," Nabile Farès writes. "Alone. Unique. Undivided. This is false. The real world is many. That's the real world. The other. Your war-obsessed world isn't the real world. *I laugh at seeing you chase after your own position. What is your independence worth if you reject plurality.*"[29]

In 1974, Farès, whose father, Abderrahmane (1911–96), had been removed from power and struck from the memory of post-Independence Algeria after having played a significant role throughout the war and Franco-Algerian negotiations, seeks in his novel *Mémoire de l'Absent* to transform through writing the practice of disappearance into a mere absence, a preliminary step in the grieving process. He shows how the father's disappearance leaves the son totally abandoned and adrift in the world, which reignites the desire to kill: "The country kills," he says.[30]

In this novel, the narrator speaks of his father's arrest and transfer to the Villa Susini (a key site of torture in Algiers controlled by the French army during the War of Liberation). In this time of a "breaking world," he writes,

> This body never stops following me, dim reflection of my disarray, so I entered it and began to wander, throughout the days that followed my father's arrest. ... As though in taking away my father they had removed my ability to understand my name, its syllables, sounds. And in this way, unable to understand myself, I had to keep my head above my delirium.[31]

In a context of systematic torture and bodily dismemberment, that his father will disappear is all but guaranteed. Abderrahmane Farès was indeed arrested and held in isolation by the Algerian authorities amid a sweeping purge of opponents of the newly "Independent" country of 1964. The overlapping of histories within History is a strange thing to behold. It is as though parts of colonialism survived its collapse in spite of Independence.

In addition to the disappearance of his own father, Farès treats the disappearance of the paternal function under colonial rule, which engenders the law of murder:

> They had gotten rid of the father, just the one they needed to get rid of so that, sooner or later, this country, this people could hope only to become terrorists. Terrorists? Everyone will clearly become one some day, since that's how it works: war, caught within the New Order, lasts forever. The New Order? Yes. The New Order, distributed between rival language, national, and religious groups.[32]

Published in 1974, this text predicted with stunning accuracy what was to happen during the Internal War of the 1990s.

4

Colonialism's Devastating Effects on Post-Independence Algeria

> No agreement is possible between the native population and the French Algerians; there is not enough space to explore this here, an entire volume would hardly be enough. In a word, I no longer believe in French Algeria. The species I belong to are monsters, mistakes of history. There will be an Algerian people that speaks Arabic, whose thoughts and dreams spring from Islam, or there will be nothing. ... The Algerian people may be wrong, but what it seems to want is to constitute a real nation, which can be for each of its sons a natural fatherland rather than an adopted one.
>
> Jean El Mouhoub Amrouche, 1955[1]

> The colonized are consumed by fear and terror.
>
> Ferhat Abbas, 1980[2]

Algerian Independence was followed by a series of unacknowledged killings, falsifications, and attempts to suppress languages and distort history and religion [*détournement des langues, de l'histoire et du religieux*]. This was all carried out by the country's new leaders. Look no further than Independence Day itself, July 3, 1962, the day a referendum was held on self-determination. Rather than choose this day, the official day of Independence became July 5, 1962, as though to erase the anniversary of the French invasion on July 5, 1830 – or maybe to make it unforgettable?

Independence fell victim to the same trap it was trying to escape, namely, the psychic, symbolic, and geographic occupation by a conquering force. Having experienced in very real and intense ways the process of confiscation over the course of 102 years led in part to an uncontrollable obsession with appropriating any form of power (political, interpersonal, linguistic, scientific, etc.). A pressing anxiety pushed politicians to become masters and owners of a language, land, and history as quickly as possible. And, in one way or another, with renewed faith in the current political power, individuals allowed themselves to become blinded by the same fantasy of possession. This time around, this also helped avoid the enormous blow occasioned by the widespread loss of language, history, mythology, land, and, above all, dignity under colonialism.

The mutilated body of the colonized and the hunger for reparation

Repeated subjection to foreign conquest and occupation likely upended the feeling of "having" for several generations. Indeed, what does "having" mean for a subject who refuses to admit the loss indelibly inscribed in its flesh and name? How can one achieve a sense of ownership if one has repeatedly been deprived of it by the Other? How do you perceive the foreigner if this same foreigner has historically been the enemy depriving you of everything, including your self? After all, what is "having" if not accepting the risk of losing what you believe to be yours, and therefore acting on this risk with others? Ownership relies on a partial fiction, since it only makes sense if it is accepted by future generations. But Algerian Independence was founded on another fiction, namely, that of believing in the existence of an ownership stripped of alterity (others, future generations, successors, etc.). This political fiction entails a very particular relationship to time: it is as though tomorrow doesn't exist. And with this doomsday scenario, zero attention is given to the future or to successive generations.

Seeking a way out of the terror that had besieged them, the victors of the War of Liberation were drawn to the illusion of recovering the power, languages, and history that were taken from them by colonialism. The opposite occurred: there was a resurgence of tyranny and citizens were stripped of their "agency." The serious losses suffered under French colonization were never processed as grief. Instead, they persisted as mere grievances and observations drowned out by a feeling of glory and triumph and the growing cult of martyrs. Many attempts were made to abolish parts of this history and to patch over the unspeakable act of destruction and the atrocities of the Internal War. As a consequence, history was severely amputated with many events disappearing from memory, leading to a crude binary between the War of Conquest and the War of Liberation. Colonialism's practice of disappearance continued unabated after Independence with the same goal: to eliminate any heterogeneous element from a unified system of power and make anything disappear that undermines the reign of "having."

The colonial practice of erasure and disappearance of people, symbolic orders, and land afflicted subjects with a pain similar to amputation: the amputee feels the pain of a "phantom member" to such an extent that its disappearance is questioned. But the goal of restoring a phantom member (language, history, and names) by way of erasure is at odds with commemoration and therefore the possibility of remembrance. In the wake of Independence, "nationalizing" language, religion, and history was a means of erasing disappearance, of making it disappear. However, that which disappeared launched an aggressive invasion. Time found itself stalled in an eternal present, unlike memories, which localize a past moment in time and space and create a narrative that must be endlessly propped up. Presence becomes unhinged from absence. The subject feels what it no longer has to the detriment of the rest of its body. Pain makes what is no longer there all the more present. The feeling of absence gives way to an acutely painful presence of what is, for its part, no longer there, while nothing can stop this pain. The awareness of absence and of what is no longer there has no impact on this self-propelled pain.

For the amputated subject, the damage is real. Its pain attests to this – it remains unforgettable. This is far from an imaginary pain. Rather, it is a case of the real surging forth at the site of disappearance. The pain of the phantom member is all-consuming and pervasive. The specificity of colonial violence doesn't allow for separation or forgetting.

And with good reason. Bodies are irreversibly marked by blank space and the violence of dispossession. For its part, post-Independence Algeria can't separate from its own phantom members (practices of disappearance, subjugation, and ruling via theft). Disappearance is the weapon of choice for colonizing spaces, psyches, and symbolic systems. This explains why the political apparatus has stressed appropriation and recovery as the means of containing and controlling the madness of those dispossessed of their bodies, history, and world. It would perhaps be more precise to talk about the madness of men possessed by disappearance and its haunting effects (see chapter 9 below).

In post-Independence Algeria, many were swayed by the illusion of recovery when what had been lost was considered to be found. This amounted to believing that the pain from the "phantom member" could only be treated by re-grafting the missing part in *the exact same way*. But what would this even mean, given that Algeria had never, properly speaking, been a nation up until this point? The whole notion of an identical graft was fueled by a fantasy that only resulted in making what had disappeared disappear once again. The goal was to avoid grafting an organ that could serve as a vehicle for an otherwise welcome and compatible form of alterity. In the imaginary, the foreigner remained cast as the enemy and, as a consequence, only the self-enclosed community (threatened by the phantom) guaranteed safety. From this sprung a persecutory reasoning that saw in difference the murder of the self and/or of the Other. The rise in conspiracy theories and their accompanying acts of persecution comes from this fraught relationship to the foreign.

The experience of disappearance leads to a painful, irreversible, internal loss that the colonized subject refuses to recognize. This is the effect of colonial trauma. Jean El Mouhoub Amrouche called attention to this state of refusal on the part of the colonized as early as 1943 in a wonderfully poetic text that serves as linguistic vehicle of bodily pain. Jugurtha, King of Numidia (present-day Algeria), who fought valiantly against the Roman occupation, is described in the following way:

> Jugurtha adapted to every situation, he was close to all of the conquerors; he spoke Punic, Latin, Greek, Arab, Spanish, Italian, French, and neglected to give his own language a written form; he loved, with the same unyielding passion, all of the gods. So it seems he would be easy to conquer. But right when conquest seemed complete, Jugurtha would rear up and escape what prided itself on being a firm grip. You're speaking to his remains, to a simulacrum, which is who is responding to you, at times agreeing with you; but his mind and soul are elsewhere, irreducible and silent, called forth

> by an inexorable voice from within which even Jugurtha thought had vanished forever. He is returning to his true fatherland, which he is entering by way of the dark door of *refusal*.[3]

Jugurtha can espouse the languages of his conquerors and praise their gods, but a large part of his self is elsewhere. Refusing subjugation at all costs, he protects his language from the colonizer by making sure it never enters into writing. And yet, returning to his "true fatherland," he brings with him this refusal directed at the Other (the occupier). And if this refusal was now directed at his own people? Isn't this one of the central problems concerning the position of the colonized? How does the self-refusal provoked by the colonizer still persist after the liberation? More than land, are the real stakes of occupation a matter of inhabiting the mental space of inhabitants for generations?

As this case becomes the scene of an internal struggle against disappearance, Amrouche shows that this process of introjection (this absorption of the outside) entails destruction and erasure. The "eternal cry" of pain turns Jugurtha the Warrior into Narcissus, whose self-obsession endures his entire life:

> Jugurtha sings what he feels when, like Narcissus leaning over the fountain, he lets out a wail *that is heard like the eternal cry of desperation from a man who has been abandoned, the plaything of omnipotent forces that have their way with him. These forces are not only external; the most dreadful, he's well aware ... come from within and, no matter what he does, they are driving him inexorably to his own end*.[4]

Writer, poet, and journalist, Amrouche was a fervent supporter of Algeria's plurality. Well before the war, he argued that Independence could only work if it embraced the land's plurality of languages and religions. It could do this by drawing on the land's history of hybridity, which goes back to the history of Numidia (the ancient Berber kingdom). Amrouche pits Jugurtha, the figure of the foreigner (an adopted child of a slave), against Narcissus, who sinks into the reflection/hole that captivates him. For Amrouche, Independence plays out at the intersection of these two figures: Jugurtha and Narcissus. In both cases, loss is at stake, but it is inflected differently in each case and ultimately leads to different outcomes.

Jugurtha: "[P]laything of omnipotent forces that have their way with him. These forces ... come from within and, no matter what he does, they are driving him inexorably to his own end." This end is not, however, the same as disappearance. It can also trigger the reinvention of "identity," as can be seen with this character who goes from being an adopted child of a slave to the Warrior King. Narcissus, on the other hand, lacking any form of alterity, disappears in his own reflection, consumed by what ends up being a hollow pit rather than a mirror. Blinded by his own obsession with his reflection, he fails to see the sky and therefore consider himself differently, like one element among many others: sky, land ... and the rest.[5] Self-obsessed, Narcissus, whose relation to the world and therefore himself lacks any foreignness, disappears into his own image. As Freud has eloquently expressed it: "Love for oneself knows only one barrier – love for others, for objects."[6] Narcissus can't agree to separate from his reflection. This is what leads to his downfall.

Reclaiming Jugurtha as an Algerian hero masks his crimes and his outsider status in regard to the royal family. This hero-worship of the man who fought against the Roman Empire distorts his image, turning him into the twin brother of Narcissus and setting the stage for fratricide (see chapter 8 below). "In order for Jugurtha to triumph over Jugurtha," Amrouche explains,

> he will need to acknowledge what he is lacking and what he needs to acquire if he wants to compete with his Western masters, rather than deck himself in borrowed plumes. ... He can't proclaim himself the equal of the Westerner by aping him or claiming the Westerner's discoveries as his own. It is not simply a matter of learning, but of inventing, creating.[7]

Since the war, Algerian literature has taken up this work of invention. However, this work only reaches a small minority. This remains a major obstacle. Without a more large-scale effort of reinvention, the social and political order will continue to be plagued by this problem. Many factors prevent this work of invention from being widely embraced. The cult of the hero, for one, masks an internal drama that remains unnamed and yet *impossible to forget*: it lacks remembrance. The myth of the hero conceals a lie, which, according to Freud, suggests that "the individual accomplished the act alone, ... whereas it was carried out by a band of brothers who rose up against the primal father."[8] The heroic son erases from the picture the brothers who helped carry out the murder. In other words, the light focused on the hero is blinding, leaving the story of his birth in the dark (the bloody relationship between brothers). In the context of Algerian history, this raises a significant question: where did the brothers of the hero go? And why are those who are still alive tearing each other apart over a legitimacy that has yet to be found? Or consider this question the other way around: what form of illegitimacy is provoking this frantic quest for legitimacy?

Colonial *hogra* and a frantic quest for legitimacy

Legitimacy is the recurring question raised by the Algerian political establishment since the emergence of national movements. The strength of political discourse and practices is measured by the claims they make for historical, religious, and linguistic legitimacy. This competition for legitimacy preceded Independence. Overthrowing those in power [*détournements de pouvoir*] has in effect been a constant of Algerian politics since the emergence of the FLN in 1954 with its *coup de force* against Messali Hadj's competing nationalist party (see the following chapter). The experience of illegitimacy is a symptom of a politics framed as an attack against colonization. In what ways is the illegitimacy felt by brothers of the Revolution tied to the disaster wrought by colonialism? Can one invent a politics that doesn't depend on refusal?

These questions are all the more urgent today, as Algerian presidents are usually depicted first and foremost as heroes of the War of Liberation. Their form of nationalism depends on unconditional love for the "fatherland." Having fought in the war gives them a sense of undisputed and unquestioned entitlement. Following this line of thinking, the fact of having been a *moudjahid* (combatant) or of coming from a family of *chahid* (martyrs) is enough to qualify them for president, or, worse, stands in place of a political platform or a need for new ideas.

From a psychoanalytic perspective, the question of legitimacy is tied to "forms of symbolic recognition," especially that of lineage. Concerning the War of Liberation, despite the active role played by women, men are the ones who are haunted by this troubling question. Looking back in history, it is clear that the question of legitimacy was a motive for many murders committed between brothers of nationalist movements before and after Independence. The wreckage left by this quest of legitimacy has been profound. Sons lacking symbolic recognition have routinely turned to war and violence, even though the Revolution cast itself as a surgical operation aiming to treat the wounds caused by colonialism. The illegitimacy of abandoned sons is a direct result of the erasures produced under colonialism: the destruction of lineages created illegitimate heirs, spurring on a frantic need to accumulate possessions as a response to the dispossession practiced under colonialism. As we've seen with the false patronyms given to men and women, offense was passed down from generation to generation.

Above and beyond the colonial *humiliation* belonging to the imaginary, *offense* and *contempt* worked on the level of the real, striking bodies at their cores. These three terms are embodied by a single word – *hogra* – which has become a refrain heard in the speech of individuals in Algeria. The feeling of *hogra* is pervasive, unrelenting, and oppressive. It orients one's relation to the self and to the Other, and, worse still, to the world. The signifier *hogra* is an archive of a past history that remains very present. The expression of *hogra* by patients undergoing treatment – not to mention countless individuals in Algerian society – makes it difficult to determine what belongs to the level of the real and the destruction experienced firsthand and what emanates from the imaginary with its set of grievances and demands. *Hogra* tells the story and, at the same time, keeps it unchanged. Consequently, as soon as the destruction caused by colonialism is brought up, we enter into a world where distinctions between various levels begin to blur. Discourse dealing with these matters quickly gives way to misunderstanding, persecution, shame, and, finally, different modalities of negation (denial, disavowal, refusal, etc.). Full-throated demands made alongside displays of ignorance continue to spread confusion and provoke war, while thinking is met with new restrictions.

The functioning of denial helps explain how these passionate responses remain detached from the will of individuals. Generational memory is troubled by denial and disavowal, which finds expression in refusal, in an excess of memory, or even in the inscription of blank spaces. Sometimes, all three of these expressions are unwittingly aligned. Denial, a powerful expression of agency, defies the passing of time: it works to reinforce a malfunctioning memory. The effects of colonialism are inscribed on a different level of the psyche than that of repression (which impacts the colonizers as much as the colonized). With denial, the possibilities of both forgetting and, its corollary, remembering are refused. The laws of non-inscription – or the inscription of blank space – obey a colonial logic that works at all times with non-rights.

It would obviously be worthwhile to square this interpretation with the lessons learned from psychoanalytic treatments concerning subjectivities and social cohesion in France. For my purposes here, I will remain focused on Algeria, where the matter of blank space is impossible to avoid. Psychoanalytic treatment in Algeria suggests that there are strong, historical ties between the rise of the disappeared, disappearance, and denial. The forces behind these phenomena work together to blot out and erase memories, creating in their place a novel, hollowed-out space rather than a dynamic and receptive surface for traces. The return of this blank space, witness-trace of what has been erased, is only brought forth and reproduced in an act of unnameable violence and, above all, through a tear in the fabric of history.

In Algeria today, the traces of colonialism are vividly present: they are seen in its architecture, in the role the French language plays, in a discourse filled with individuals' grievances and vindictive feelings, and so on. And the very abundance of traces bars subjects from participating in a collective reconstruction of history and languages. It also prevents them from sharing in the feeling of having any political worth. This excess of visible memory stands in place of a hollowed-out memory, foreclosing the possibility of rewriting and severely limiting interpretation. This is all the more true, given that the totalizing power of the political order establishes and reinforces historical memory as a source of legitimacy. This practice ensures that denial exercises control over the subjective effects of history and leaves the question of legitimacy in abeyance.

The "orphaning" effect of colonialism and its impact

In his writing produced between 1943 and 1961, Jean El Mouhoub Amrouche lucidly illuminates the obscure feeling of illegitimacy, which has reigned since the beginning of colonization. Independence failed to liberate the sons leading the Revolution from a profound struggle with denial. Amrouche traces the origin of denial back to a matter of filial lineage:

> The colonized subject is denied his lineage, which originates in myth and history and continues on up to the moment when a free people becomes an enslaved population. … The colonized subject is denied, and *must feel defeated among his ancestors. The consequences of this defeat will last indefinitely*, spanning both the past and the future: *he is thus unmoored from both past and future generations.*[9]

Kateb Yacine and Nabile Farès both framed fathers as disappeared figures who left sons to grapple with a collapsing lineage. The place of fathers and the disavowal of their role under colonial rule deserves serious study outside of the accounts provided by literature. Indeed, there was never a founding figure who commanded widespread respect (with the exception of the Emir Abd El-Kader, who took refuge in Syria after France's colonial invasion, and of Messali Hadj, who was shunned by his own people); what's more, the Algerian Revolution had always been presented as springing from an alliance of brothers. The signifier "brother" would come to occupy a very important place in relation to the Revolution, downplaying the brothers' highly coveted place as elected sons. Unless the brothers leading the Revolution are considered fatherless sons. In this case, the term "brother" also shifts attention away from the fact that these sons couldn't properly position themselves as sons (sons of/*ibn* or *ben*).

The sons of disappeared, unburied fathers, as we've seen, are also grappling with the disappearance of their original patronyms. These erasures reduce the "*indigènes*" to nameless bodies, or bodies with fictitious names. Amrouche points out that the children born under colonialism were suspended in a *no man's land*[10] of symbolic heritage. The symbolic world of their parents had collapsed, leaving them with nothing to cling to, while French education promised little by way of linguistic or historical belonging, as they remained decidedly outside the French imaginary. Indeed, on all these grounds, they remained excluded. Amrouche notes that adults who lived through this as children are afflicted with a sense of illegitimacy. The colonized subject, he says,

> is not a legitimate heir. And, as a consequence, he is a sort of bastard. Bastard by necessity, since the legitimate heir, the heir by right, only exists unconsciously without getting to experience the privilege of inheritance. The bastard, for his part, barred from inheritance, has to earn it on his own; assuming the position of heir by dint of his

own effort, he is able to recognize and fully appreciate the value of inheritance.[11]

This effort to reappropriate heritage and therefore find one's place in history was at the center of a war that cast itself as revolutionary. However, the feeling of illegitimacy cannot be settled by recovering what no longer exists. This reality led to an endless internal war waged over the right to legitimacy. Unable "to be," the next best thing is "to have" in order to fill the hollow space left by an absence of existence. *Hogra* has its real origin in the dispossession of names, languages, and lineage under colonialism. This is what caused its aggressive spread. It also furnished the post-Independence illusion of total recovery, which sought to regain, almost item by item, all that had been lost. Unfortunately, what had been lost no longer existed. How do you recover the irrecoverable? Doesn't that invite a form of madness?

Indeed, the more a subject chases the irrecoverable, the more it falls into a mad frenzy of "having." Through this, it is seeking to escape its state of melancholy occasioned by significant loss. This proves to be all the more challenging since changing the father's name alters the patriarchal functioning of the social bond. "When [the colonized subject] develops self-awareness and tries to find out who he is," Amrouche remarks, "what he first discovers is his own uprooting."[12] Indeed, the feeling of *hogra* springs from real damage at its origin. But the work of erasure severed this feeling from a concrete reference, forcing it to exist in the imaginary as a mere grievance. The site of damage has fallen into oblivion or been erased. Existing in the imaginary, *hogra* seeks to address its real causes, but these are buried at an unknown site. The hollow core of *being* – the mark of colonialism's damage – is either erased or denied while *having* assumes increasing importance: in Algeria this occurs through the act of subjects recovering what is "rightfully theirs" from the colonizers; and, in France, through a discourse promoting the "benefits of colonization." France's history of Algeria, which is starting to disappear from memory, still wields a lot of power.

An interesting detail of this history illustrates the crushing feeling of destitution underlying the logic of "having." In the immediate aftermath of Independence, when Europeans were fleeing the land *en masse*, the goods they had possessed for generations were deemed "forfeited assets." A government directive then led to a rush on these "assets." Vacated spaces were to be filled at once. This frantic rush embodies the essence of the drama: the experience of an irreversible expropriation and a feverish turn toward appropriation as a means of escaping dispossession. Subjects were thus urged to regain a "self" as quickly as possible, even if they never possessed a "proper self." This rush to fill vacated land shows how the fantasy of recovery took the place of the much-needed work of clearing out interior (psychic) and exterior spaces. Rather than merely identifying with the colonizer, he was being usurped.

This role-playing is provoked by the fear of returning to one's "self" and failing to recognize it. With the disappearance of its distinguishing features, colonialism has made it unrecognizable. The pressing need to fill spaces as quickly as possible is therefore tied to the deep-seated and persistent anxiety of being exposed and stripped bare, a feeling that has remained unchanged in spite of the liberation. Reappropriation does little to appease the horror and damage occasioned by an immense loss. The best the subject can to do to address this is to pose as a new colonizer.

Albert Memmi sees this as the tragic fate of the colonized, who is forever caught between two problematic identifications. To get out of this bind, he must rid himself of both his colonizer and colonized self. What was left of his "own" self after the colonial disaster was driven out by these two identities. This sort of subjective experience is without precedent. After colonialism robbed him of a self, the subject now has to get rid of the colonized being he has become. For Memmi, the colonized subject must abandoned these two identities (colonizer/colonized) and reinvent a new way of being in the world: "To live, the colonized needs to do away with colonization. To become a man, he must do away with the colonized being that he has become. [T]he European must annihilate the colonizer within himself, ... this creature of oppression and want."[13]

With the individual failing to separate from his or her identity as a colonized subject, a mounting crisis of subjectivity threatens to wreak havoc, hollowing him or her out from within. Posing as a new colonizer is but a trick to hide the existence of a shattered self. In short, it is better to pose as a faux-colonizer than to live in an abolished dwelling. Failing to find him- or herself, the former colonized subject takes him- or herself for the Other. This fact is central to understanding how loss suffered under colonialism becomes irreversible. No recovery is possible.

Feelings of shame, contempt, humiliation, and offense plague an existence one cannot easily renounce. Amrouche asks how one can care for what has been stripped of all humanity, for an individual who has become an object at the mercy of the Other. There is little likelihood, he points out, of finding a treatment that could function collectively for a deeply ingrained *hogra*: "How can you get over humiliation? Especially the sort of unsparing contempt and humiliation felt in this situation. This eats away at the very core of individuals, destroying that which makes them human."[14]

With new patronyms assigned in the 1880s, bodies were unmoored from their names and, by extension, lineages. This provoked a feeling of illegitimacy as the loss of heritage translated into a loss of self. As we've seen, the disappearance of names cleared the way for the disappearance of bodies. And with unreliable names, there was no clear way to keep a tally of the dead. The living found themselves on the same level as the dead. The impacts of all this were felt by future generations in a context of unspeakable violence: "It is important to understand," writes Amrouche,

> that the first condition of existence is to have your own name, one that hasn't been stolen, usurped, or imposed. That some individuals, in atypical or exceptional cases, who have been uprooted from their own racial history, occasionally succeed in establishing new roots in the body of an adopted nation is perfectly conceivable. Just like it is conceivable that some émigrés come to forget their country of origin which they usually fled for good reasons. But to subject an entire population to assimilation entails systematically destroying what defines it as a population, which, properly speaking, is called genocide.[15]

Further distortion of patronyms

This discussion is all the more pertinent after many patronyms were once again changed in Algeria following transcription "errors" during the transition to biometric passports in 2009. The fictional nature of Algerian patronyms was fully revealed on this occasion as they moved from one language (French) to another (Arabic), re-enacting the destruction of names during the colonial period. Names inherited during colonization – as a means of controlling the population and land – became almost unrecognizable once they were re-transcribed into Arabic. This is because the Algerian civil registry continued to rely on the re-transcription policies put in place by the colonial administration, which is to say that it dismissed lineage.

Do these Arabized names betray the longing for the disappeared names of ancestors? Or do they betray the hate for a fictive, inherited name? In either case, countless Algerians have experienced the persistence of a mutilated name many years after Independence. Retrieving original names at the civil registry entailed costly and burdensome legal proceedings. Thanks to administrative officials, the new Arabized patronyms are a brilliant illustration of the untranslatability of a past history, which, inscribed on bodies, remains forever present. For example, within the same Lazali family, such as mine, some bear the name given by the French administration, and others have been given the name Laâzali or Lazli – in Arabic, *aâ* is an entirely different letter (pronounced *'ayn*) as opposed to *a* (pronounced *aleph*).

The new patronym is entirely unrelated to the original (itself a false name given by the colonial administration). The patronyms once again modified by the Algerian administration reinforce, in both their written and spoken forms, the process of disaffiliation and renaming without regard to lineage. Other patronyms suffered even further disfiguration from the addition or removal of several letters. This mutilation of names clearly reveals how colonialism is still at work in the self-enclosed community.

This forced almost the entire population to provide proof of their lineage and to seek out the transcription of their names during the colonial period. But, as we've already seen, even during that time, patronyms no longer reflected traditional naming practices, with siblings divided by their patronyms. The patronym had lost its status as a symbol and identifying feature of a family/tribal body, and therefore of the land that belonged to the family group. This type of mutilation sought to undermine heritage and belonging. Although this history has been ignored, the practice has only become reinforced in post-Independence Algeria.

We have gone from a deliberate act stemming from a policing tactic to transform patronyms during the colonial era to so-called typographic "errors" – both of which reveal the fictitious status of the name under colonial rule. The impossibility of transcribing names in post-Independence Algeria is in fact part and parcel of the colonial process, for the patronym bears witness to an impossible heritage. Its function has become one of disruption rather than unification, recognition, and identification. The mutilation of patronyms in post-Independence Algeria is a living memory of the destruction of names under colonization. History is being re-enacted, as though the creation and disfigurement of names by the Algerian administration offered a way of refusing to forget what was carried out by the colonial administration. Are we forever condemned to be children without a lineage?

In a similar manner, renaming streets in big cities, from the names assigned during the colonial period (therefore French streets) to the names of martyrs of the War of Liberation, added more confusion to the process of naming and organizing space. This effort has been plagued by untranslatability, dividing those (older generations) who know the streets by their French names and those (newer generations) who, unfamiliar with the older designations, use their Arab names. This constant state of disorientation is a symptom of *unbridgeable divisions*.

Divested of a name: a form of colonial murder

Albert Camus' novel *L'Étranger* masterfully illustrates how the "*indigène*" is robbed of his name. He is the "Arab" and "the stranger," caught between two totalizing positions: a self stripped to its core (reduced and divested of humanity) and a radical Other (the stranger). His ethnic belonging (Arab, Kabyle, Tuareg, Mozabit, etc.) and his history do not matter. The absence of a name authorizes murder, and at the same time this absence is a form of murder. Camus was torn between two spellings for the name "Meursault": "Mersault" (*mer/mère*, *saut/sot*, etc.) and "Meursault" (*meurs/meurt* in the first-, second-, or third-person singular and *saut/sot*).[16] In this way, the signifier "murder" plays on many levels: the real and symbolic, the absence of a name (the "Arab")/presence of murder in the name ("Meursault"), and finally as an act of murder itself. Camus demonstrates brilliantly that murder runs through coloniality on a level at once explicit and implicit, written and unwritten, legible and illegible.

Many murders are interwoven throughout *L'Étranger*. On the one hand, it is inscribed in the name;[17] on the other, the Arab's absence of a name reduces the subject to an anonymous, worthless body (the "Arab"). Murder in this case is a logical consequence of being divested of a name. From an unnamed "Arab" subject we arrive at the unnameability of colonial murder. Through the name's status (absent for the "Arab"/inscribed in "Meursault"), Camus highlights the political stakes of colonialism's position outside the law. There are no nameless bodies, and therefore the absence of a name already indicates the murder of the subject. The murder ("*meurs/meurt*") within the name "Meursault" targets the nameless. *Murder strikes the "I" and the "He," which is to say, the subject and the Other.* Are both colonizer and colonized forever doomed to be bound together?

L'Étranger speaks to the disappearance of alterity under colonialism. This text sheds light on the colonial logic according to which the absence of a name already implies the murder of the subject. Amrouche describes colonialism as an "operation of uprooting":

> This operation of uprooting begins in primary school, at the earliest stage of education, in the purest and most naïve intention to civilize. The colonized child has no parents, he has no ancestors, the country where he was born has no history, no great men with exemplary stories and lives to fill his dreams or inspire him to follow their model. He has progenitors whom he is taught to love and respect for who they are. But they can't serve as guides for him. And as he sees himself as superior to them, his filial affection is complicated by a whole mess of feelings: indulgence as well as rather aggressive pity-contempt-resentment. For it isn't long before he feels within the burn of humiliation and the private disaster caused by his imperfection.[18]

In another article, he elaborates on this idea, saying:

> The subject comes to contest his own identity, he no longer knows who he is as he becomes accustomed to his fragmented state, to a spiritual and ontological tension he vigilantly observes. *But this tension threatens to destroy him because he has no name. Now, no being in the world, no human being in any case, can do without a legitimate name, a name he recognizes in himself and a name people rightfully recognize him by.*[19]

Amrouche offers to care for these damaged, nameless bodies by "easing" them back into their mother tongues. This involves reinvigorating the maternal languages whose *archi-trace* is stored and protected in the body. It is worth noting that unnatural, imposed languages work in the other direction, creating a distance that is part and parcel of the colonial process. In a 1960 text entitled "Colonization and Language," Amrouche pushes this idea even further:

> We must first of all guarantee that this man, in the depths of his being, has a strong ontological foundation, which is to say that he has a right and a direct path to his heritage via the possession of a language, the same language that makes him, since we all owe our existence to language. This is the case both on the level of conscious memory and, on a much deeper level, with involuntary memory and archetypes. For *the language in which he will take shape and which has shaped him* must be for him not merely a collection of forms or a relatively rich vocabulary which he may use, or fail to use, to his advantage as he understands its rules. The words of this language must resonate for him on every level, and its whole range of meanings, the semantic depths of this language must be felt by him in the depths of his being. Words should in a certain manner take shape within his being and not be merely just what his mind, or what his memory, consciously elects and uses. This is what should be the national language.[20]

Amrouche calls for re-establishing the lost connection between words and things, names and places, language and history. According to him, the aim of liberation should be to establish the "national language" as a site of interpersonal exchange. This goal ran aground with Modern Standard Arabic, pushing languages such as Algerian Arabic and Tamazight into greater obscurity. These languages, despised and forsaken by disciplinary knowledge, are confined to the private sphere, where they fail to put up any resistance to the political order, thereby bolstering the reign of prejudice.

In the days following Independence, the new political power had the right intuition: to repair the effects of colonialism through language, but the craze for possessing and "having" drove it toward nationalizing the language, which is to say, appropriating it. This fueled the fantasy of *being a master of language rather than a product of it*. The right intuition transformed into a totalitarian ideology: to turn language into a divine, inalterable body-armor.

Manufacturing erasure and denial under colonialism

Moving beyond Algeria, it is worth considering how the mechanism of denial silently penetrates History, especially in its relation to murder and destruction. The erasure of traces leads psychoanalysts down the paths of denial. This mechanism is crucial for understanding the relation between the individual and the community. Deciphering this territory is complicated by denial's express purpose to erase traces and occasion disappearance. The psychoanalyst Claude Rabant explains denial in these terms:

> As opposed to repression, denial doesn't work by leaving marks, but by erasing them from discourse. ... It tends toward self-erasure by undoing textual space altogether. It negates the very existence of a foreign site, the presence of a displacement or a lapse in time. It eliminates an outside, an alterity, a foreignness, which are all experienced as crimes.[21]

Totalitarian policies can be observed gaining favor throughout the world. Perfectly aligned with capitalist discourse and the spirit of globalization, both of which adopt exclusionary practices, totalitarianism is rapidly spreading. Its policies seem especially well suited for keeping denial at the center of the collective body. In order to accomplish this, a totalizing logic seeks to erase difference entirely. Indeed, these regimes manufacture erasure on a large scale. This could also be put the other way around: these regimes are manufactured by a collective denial of History. The traditional opposition between the governing class and the governed class would thus give way to a regime of governance founded on collective denial. The disappearance of traces is its principal strategy, the way it makes history, while a *blanked-out* memory provides its foundation. This gives rise to a constant feeling of insecurity, one that is tied to an internal vulnerability. Hence the obsession with establishing a definitive origin where one can repeatedly find one's bearings.

The paths of disengagement have disappeared, which translates into an experience of extreme saturation. Getting out of this situation comes with a hefty price and agreeing to attendant losses, which are certainly irrecoverable. But it is this loss that is refused, because the colonized subject believes everything he had was already taken by the colonial Other. After Independence and the departure of this colonial Other, the rush to recover what was lost gave him the illusion of finally feeling well provisioned, or, worse still, fulfilled. But a deep-seated feeling of injury continued to hollow subjects out from the inside and to plague both the private sphere and the social bond with violence. Reflecting on the positions of the colonizer and the colonized, Albert Memmi gives a striking image of this feeling of saturation, which expresses the refusal of loss and the potential experience of a very real and deadly pleasure: "The decolonized is thus led to engage in a zigzag march between an increasingly frayed national present and a distant utopian future. *Having had his fill of the pleasures of independence, he is barely moved by the signs and symbols of sovereignty.*"[22]

The "sublevel life" (Nabile Farès) that the subjects lead, despite the fulfillment mentioned by Albert Memmi, serves to create a trace of the injury they experienced and, above all, to commemorate it. This also allows them to present themselves as subjects of exception whose plunge into melancholy is a matter of fate. The promise of relief offered by grief is taken from them as they succumb to a *frantic state of uninterrupted melancholy*. In Algeria today, it is not uncommon to hear someone say, "but it's not like that here," especially when scholars are visiting from abroad. The status of legal exception, as under the *code de l'indigénat*, has paradoxically become widespread in the postcolonial era, while no one seems aware of its colonial origins.

The logic of injury seeks to expand murder and destruction within the social arena. It also works to suspend time: "The injured subject," writes the psychoanalyst Paul-Laurent Assoun,

> turning to a past injury in order to avoid confronting a coming one, is stuck in the present, which remains a stumbling block it can't overcome. Finding a dialectic in its history is no longer a question. Which is why, judging the current point to be the outer limit of concession, it refuses to go beyond it. The dreadfulness of the past justifies a negated future, ... where as a sort of exception it stands outside the universal human condition.[23]

From colonial trauma to social trauma

After Independence, colonialism's practice of exclusion, experienced on the level of the real, was translated into a legal exception, which is to say, an exception to universal principles. This amounted to no more than a reprise, and extension, of the first exclusion. The subject of a so-called "Independent" country found itself, to its dismay, caught in a system of exclusion of its own making. This is the logic of trauma, which has a profound impact on subjectivities and their expression in the social arena. The injury caused by various colonial abuses takes root in the present as "social trauma." The past is an enduring present. It becomes hard to distinguish what belongs to the subject, to the public sphere, and to the political sphere.

Existing as exceptions to the law during colonization, there was still a struggle after Independence to shake off this "exceptional" status. The practice of exclusion continues unabated and one's status as existing "outside of …" provokes a terrible feeling of shame. The fate of injury depends on this demonic figure of reproduction, which repeatedly condemns and confines one to an ashamed self. As Amrouche has made clear,

> For the "*indigène*," disdain and humiliation speak to the very quality of his being "*indigène*" (so do the insults *bicot*, *raton*, *tronc de figuier* that are directed at him), they are attached to his being like a physical trait or chemical property is attached to a body; they affect the totality of his historical existence, reaching the past and the future of an entire people.[24]

Reducing the subject in this way to a shameful body creates an identity founded on exclusion (from one's self, the Other, the world). It forces the subject to only recognize itself in the image created for it by the Other through a real or imagined gaze. Shame is the sign of a radical nudity. It indicates that the subject is struggling with the shame of existing as no more than a naked and nameless body. As Assoun writes, "To be ashamed is to feel identified with one's self over and over again. … The deep cause of my shame … *is my inability to be other than myself*."[25] The ashamed subject is reduced to its own self, exposed and lacking … Which explains why it is always dressing itself in multiple layers (ideological, religious, moral), to the point of suffocation. It also has to buttress up its ideals, which hide its misery. It becomes easier to park in shame than to separate from it. Shame can lead to patching over the loss of being with an excess of having, which fuels further demands. Along the way, the ashamed subject finds hate, both for its self and for the Other. This affect, derived from colonialism's criminal enterprise, continues to wreak havoc in a variety of paradoxical forms. The persecuted thus becomes the persecutor.

For this explosion of hate is too much to bear for both the individual and the larger public. Unacknowledged and unidentified, it has no specific target. It consumes the subject and devours its "book of flesh." Hungry for revenge, it remains insatiable. Shame slowly consumes the private sphere and leaves behind only the discarded "cores" of bodies. By spreading to others, hate promises to escape one's self, and, therefore, act as a check on shame. But the difference between hate and shame vanishes when they find themselves in pursuit of the same goal: namely, to conflate the self and other to the point of indistinguishability. Each of these affects casts itself as a remedy for the other, whereas they collectively work to achieve the same objective: disappearance. Thus, the subject has a choice between disappearing into hate or disappearing into shame.

Seventy years after Camus published *L'Étranger*, Kamel Daoud, in his novel *The Meursault Investigation* (2015), re-establishes the site of interpersonal exchange which had been broken by the colonial system and which was absent from Camus' novel.[26] Daoud creates a new space for this by giving a name, a lineage, and a history to the "Arab." This gives colonial hatred a clear form and expression. In this way, Daoud also brings visibility to the silent blank spaces in Camus' novel – those sites of vast destruction. The novel's invisible writing is thrown into relief, allowing the injured party in the text to tell his own story and escape the demonic curse of a shameful fate. The crushing weight of murder becomes, in Daoud's writing, an act of creation for something that had no history. If Camus makes colonialism's law of murder extraordinarily clear, Daoud examines its systemic organization and the impact this has on subjectivities and policies. It is indeed a *counter-investigation* [the French title of the novel is *Meursault, contre-enquête*] of the primary murder committed by colonialism: the absence of a name. To read *L'Étranger* today without this counter-investigation thrusts the reader into the illusion that the real murder is secondary to this story. For the primary murder is being dispossessed of a name. More than half a century after Independence, Daoud's act of restoring the name was long overdue. This is the role and purpose of Algerian writers since the birth of Francophone Algerian literature in the 1940s and 1950s: to find a manner of engaging a wounded public and resisting the political order.

In post-Independence Algeria, colonial trauma slowly turned into "social trauma." This entailed translating into the social order what appeared in the subject as blank space, which, as such, remained unreadable. This unfolds on both a private (*intime*) and "exteriorly private," or *extime*,[27] stage of the public. As my clinical experience in Algeria over the past two decades has shown me, this social trauma overwhelms and dissolves the subject within the larger public. The impact of denial, without doubt one of the principal mechanisms of colonization, has been twofold: it has affected individual subjects by falsifying their connections to the symbolic order (genealogies, language, history) and has affected the particular make-up of the colony regarding its relation to the law and rights. Hence a divided symbolic order, reinforced by an erratic legal system which protects some (the colonizers) but not others (the colonized, held as exceptions to the system). The subject is thus barred by the colonial system, left "outside of … ." This has led to psychic responses that are distinct from ordinary neuroses. With a political agenda predicated on eradicating all forms of alterity, coloniality has inflamed hatred by seeking to preserve the One by killing the Other. To what extent does the rise of "nationalism" in Algeria coincide with the barring of alterity? And what impact has colonialism's negation of the paternal function had on contemporary politics? Answering these questions requires a closer look at how the phenomenon of "fratricide" has shaped Algerian history.

5

Fratricide: The Dark Side of the Political Order

The psychic force of hatred must be greater than we think.

Sigmund Freud, 1901[1]

The very emphasis of the commandment: Thou shalt not kill, makes it certain that we are descended from an endlessly long chain of generations of murderers, whose love of murder was in their blood as it is perhaps also in ours.

Sigmund Freud, 1915[2]

To better understand the role played by the colonial system – founded on the erasure of alterity and the active killing of the Other – in the rise of fratricide beginning in the 1950s and continuing well after Independence, one must go back to the 1920s. Without providing an exhaustive account, it is worth recalling the historical conditions surrounding the emergence of various liberation movements in Algeria at that time. Paradoxically, metropolitan France was also welcoming Algerian immigrants, which illustrates the split within French politics between its colonial side and its republican side. Mohammed Harbi called attention to this in his *Mémoires*: "There were certainly two Frances, and one of them had deeply penetrated my own history in order to denounce the other all the more forcefully, ... whose repressive, colonial face was appalling."[3] The repercussions of colonial violence, both during the War of Liberation and in post-Independence Algeria, is a subject to which I will return below.

The emergence of Algerian nationalist movements in the 1920s

During colonization, a growing Algerian immigrant population was settling in France, made up for the most part of men who were fleeing the terror and misery of the colony on their own to find work in factories. Back home, these same men were also facing tragic conditions, called upon to murder either their own or the Other. Since 1912, with obligatory military service within the French army (since the "*indigènes*" had neither their own state nor their own army), they found themselves forced to fight their "brothers" in their own country. This alone was enough reason to go to France in order to avoid the fratricide orchestrated by the colonial administration, only to return to join the ranks of the *Front de Libération Nationale* (FLN) in 1954. Among these factors, living in the metropole offered another clear advantage: people were relatively free to congregate as they pleased and build a national movement. A stark contrast was thus felt by immigrant men between the rights they had in the colony versus those they had in the metropole.

Beginning in 1956, the war had shifted to the metropole, where numerous attacks were carried out and where the nationalist movement was met with violent repression. The massacre of October 17, 1961, in Paris following a peaceful protest led by Algerian immigrants demanding recognition of their country's independence was a key moment in this history. The division between the colony and the Republic proved once again to be illusory: surveillance, imprisonment, and executions were coming to define both sides of the Republic. Other places in France were able to maintain a safe distance from the colony's totalitarian regime. Relative freedom therefore still existed for young workers in the metropole as opposed to those living in the colony.

The first Algerian nationalist movement had thus emerged in the context of immigration. The *Étoile Nord-Africaine* (ENA) was founded in 1926 by Messali Hadj (1898–1974) and a few other exiles in France. This movement was headed by Emir Khaled, grandson of Emir Abdelkader, a widely recognized fighter who resisted France's invasion during the War of Conquest. Dissolved by the government in 1929, the movement reappeared under the name *Parti du Peuple Algérien* (PPA) in 1937, then became the *Mouvement pour le Triomphe des Libertés Démocratiques* (MTLD) in 1946, and finally the *Mouvement National Algérien* (MNA), which fought for Algerian independence. Messali Hadj, considered by historians as the father of Algerian nationalism, led each of these political organizations.

In colonial Algeria, the Association of Algerian Muslim Ulamas (scholars) was created by Abdelhamid Ben Badis (1889–1940), an Arabic scholar who was inspired by Mohammed Abdou (1849–1905), himself an Egyptian Mufti and reformer who had pushed to adapt Islam to the modern world. Islam and Arabic were at the center of this organization's political goals. Its slogan was: "Islam is my religion, Arabic my language and Algeria my fatherland" (*El islam dini, el laura el arabia; laurati wa El djazair; watani*). This organization should be situated within a broader global movement emerging across many Arab countries: the *Nahda*, or Arab renaissance. The *Ulamas* were driven by the political aim of establishing Arabic and Islam as familiar references. The hope was that this would allow them to re-establish their place in History and in the world, and by having Algeria recognized as an Arab and Muslim country, they would regain their dignity and status among other Arab countries. Colonialism's push for establishing an inward-looking identity also provoked the desire to form a community of belonging that would allow individuals to reclaim their dignity. This explains why Islam played such an important role in rejecting subjugation. At the beginning of colonialism all the way up to Independence, Islam was used to seek recognition (among Arab countries) and then potentially establish one's place in the world (at least the Arab-Muslim world) and in History (by *jihad*).

This throws into sharp relief the feeling of living in exclusion from the world. The desire to be included is a direct response to the feeling of exclusion caused by colonialism. The use of Islam to build a "national" sentiment thus had a paradoxical effect: at times it opened up to the Other and the world, re-establishing an experience of fraternity wiped out by colonization, and inviting an experience of legitimacy; at other times, however, it closed itself off, distorting religion [*détournement du religieux*] to serve its own political purposes, leaning on the authority of its supposed divine origin.[4] Islam was also instrumentalized in the creation of the hero figure, the martyr of the devout fatherland (*shahid*). As Mohammed Harbi has argued: "Islam stood in place of the state before becoming its soul."[5] Alas, the history of contemporary Algeria is dominated by the story of religion's triumph over politics. The other story, the one that frames religion as a source of knowledge, history, and alterity, is buried by the political order.

The War of Liberation and an impossible fraternity

The tragic political fate suffered by the father of Algerian nationalism supports this dominant story. The official narrative of Algerian Independence remembers Messali Hadj as the patriarch (*zaïm*) of the national cause whose hunger for power led to his downfall and usurpation by his sons who led the Revolution. Despite Independence, he wasn't granted Algerian nationality until 1965, and his requests for a passport were denied many times by the young state. It wasn't until shortly before his death in 1974 that he received one.[6] Thus, the father of "Algerian nationalism" would only become Algerian on his deathbed.

His place in this story reveals the hidden side of the "national" liberation struggle.[7] The parricide of Messali Hadj that occurred on a symbolic level was part of the same logic that would drive the brothers of the Revolution – made to feel illegitimate under the colonial regime – toward an insular community and fratricide: colonial practices resurfaced among liberation movements without anyone acknowledging that this was merely a re-enactment of history. Far from belittling the immense achievement of men and women who fought for liberation, this is simply a reminder that the history of colonialism didn't end there. That this history is absent from most accounts despite being repeatedly deployed by those in power even today is a telling reminder of the work that still needs to be done. Why did Messali Hadj's status as a founder of nationalism become some sort of political taboo? To what extent is today's politics shaped by a specific narrative of colonialism?

For the young militants of the FLN, Messali Hadj, who was born in 1898, belonged to the generation of fathers. And Messali's followers were deemed "interior enemies" by those who, beginning in 1954, became part of the FLN – and vice versa. The enemy looms as much from within as from without. The same "national" cause was being sought between two warring factions during the War of Liberation. The FLN, a child in many ways of Messali, was formed by removing him from power. The ancestor thus became the enemy. Each faction sought to eliminate the other and claim for itself the title of "Father of the Revolution": "What the FLN could not bear," recounts the historian Gilbert Meynier,

> was the territorial dispute waged by a rival organization, one that shared the same set of values but went by a different name. ... They shared the same political positions, as their unwavering goal remained the same: Independence. The only difference was over the means of the struggle, in the FLN's uncompromising position in favor of armed struggle. As Mohammed Harbi has made clear, within the specific environment of warring brotherhoods, "the absence of a clear political foundation personalizes conflicts."[8]

At the very outset of the War of Liberation, the executions ordered by the leaders of the *Armée de Libération Nationale* (ALN, military branch of the FLN) targeted two groups: first and foremost, the external enemy (the colonizers), but then also the "internal enemy" (the Messalistes militants). Later on, driven by the notion of purity and feelings of extreme distrust, some FLN fighters would also be labeled internal enemies. Thousands would become victims of attacks fueled by this obsession with a suspected enemy, a sense of paranoia in large part stoked by the "psychological" tactics deployed by the French army. A war between "us" and "them" was being waged on multiple fronts: between Algerian fighters and the French army, between FLN and MNA (Messalistes) fighters, and then among FLN fighters themselves. Henceforth, the original division between "us" and "them" fell apart: the war against a foreign enemy grew into a *war waged from within.*

The French army knew how to exploit this fratricidal tension, with the hope that these internal attacks would hasten the collapse of the nationalist movements. It did this principally through infiltration and the spreading of false information. This is once again a question of deliberately confusing the real and the imaginary – a defining feature of colonialism. Nowhere is this perhaps better illustrated than in the terrible "*bleuite*" affair. In 1957, the French Secret Service (SDECE) had undercover Algerian agents called "*bleus*," or "blues" (owing to the color of their work clothes), pose as brothers of the nationalist movement in order to infiltrate it. Suspicion spread rapidly and people were haunted by the question "who is who?" Thousands of high-ranking fighters within the ALN were executed as a result of this, which was a significant blow to its organization.[9] This war strategy deployed by the colonial power spread suspicion and supported conspiracy theories. Countless fighters accused of betrayal were tortured and executed by members of the FLN. The confusion between enemy and ally was thus almost complete, one that continued to reign long after this war, as we will see with the Internal War of the 1990s. This spreading of confusion made friends into close enemies.[10] The slightest internal criticism or disagreement was deemed treasonous and a sign of complicity with the enemy.

Jean El Mouhoub Amrouche's journal contains a poignant account of this war waged on one's own and the expulsion of difference within one's self. This man with two first names, one Kabyle Muslim and the other tied to his parents' conversion to Christianity under colonization, wrote a letter in 1946 addressed to his self's other:

> From El Mouhoub to Jean. My dear friend, if I've decided to write you, it is because I know, in fact, it's useless: for forty years now we've inhabited the same body, we've drunk from the same sources, we're made of the same light and shadows, we've suffered the same ills: but for the past twenty-five years we've confronted the same problems and we've unsuccessfully tried to become one. A single body, a single soul and, within this same body and this same soul, an unacknowledged gulf bigger than the sea has held us apart.

Ten years later, in 1956, he returned to this idea:

> For the past eighteen months, men die, men kill. These men are my brothers. My brothers are dying. My name is El Mouhoub, son of Belkacem, grandson of Ahmed, great-grandson of Ahcène. My name is also, indivisibly, Jean, son of Antoine. And each day, El Mouhoub hunts down Jean and kills him. And each day, Jean hunts down El Mouhoub and kills him. If my name were only El Mouhoub, it would probably be easy. I would embrace the cause of all the sons of Ahmed and Ali, I would espouse their reasoning, and I would speak unambiguously in support of them. If my name were only Jean, it would also probably be easy. I would adopt the reasoning of all the French who seek to eliminate the sons of Ahmed, and I would be able to speak just as unambiguously in support of this. But I am Jean and I am El Mouhoub. Both live in the same and only person. And their ways of thinking are at odds. Between the two, there is an insurmountable distance.[11]

Having suffered a premature death in April 1962 in the days following Independence, this major writer made the effects of the Internal War legible for future generations by insisting on its *unbridgeable divisions*. His writing provides insight that surpasses Jean and El Mouhoub and reaches further than any question dealing with the colonizer/colonized label. It is found in this internal double that hunts down and kills its identical other. Subjective space becomes the site of an internal war waged over the self's irremediable difference, which suffers no small amount of destruction in the process. This battle to the death between the psyche's two sides gives rise to an overwhelming feeling of melancholy. The subject's internal other, which forms an integral part of its self, is eventually killed off. The political sphere unwittingly took over the care and maintenance of this body struck with alterity. The lack of acknowledgment by the colonial Other (of its lineage, its languages, and its history) and then by one's self (post-Independence) created an experience of illegitimacy. Subjects are still grappling with this absence, or blank space, of symbolic recognition to the point of being banished from being the children of their own languages and histories, which are decidedly plural and diverse.

From parricide to fratricide

How to make sense of an expanding war where inside and outside are blurred? And what can be said about this haunting obsession with the enemy? Who benefits from excluding the Internal War (between nationalist militants) from accounts and histories of Algeria when it is still shaping contemporary politics? The role of the Internal War remains to this day sealed in secrecy and *blanked out* from the official narrative. Seriously undermining the heroic narrative of revolutionary brothers, it has been foreclosed from first-hand accounts, stories, and testimonies.

Born in part from the divisions created under colonial rule, this "war within the war" cannot be explained by colonialism alone. France was plagued by the same problem concerning the French military and the OAS (*Organisation Armée Secrète*) in the months preceding Independence.[12] In addition to this, a brutal power struggle was being waged between the military leaders of the border army and those of the various *wilayas* (regions), resulting in the death of more than a thousand people between July and August 1962. In response, people took to the streets to call for a stop to these massacres between warring former nationalist fighters, chanting "*Sab'aa snin, barakat*" ("Seven years of war is enough"). The seeds for a civil war were being sown in Algeria, and shortly before this in France too, as the OAS threatened to topple the dominant political power. The separation from coloniality created the conditions for civil war.

The reigning political system is organized according to a logic of elimination. It is a matter of gaining complete control (over power relations, interpersonal exchange, governance). To achieve this, there has been an endless series of coups, forced removals from power (physical or symbolic), and attempts to topple leaders, which found renewed vigor after Ramdane Abane, the leader of the FLN, was taken down in 1957 (a point to which I'll return). Indeed, as the historian Gilbert Meynier has noted, "over the course of the whole war, Algerians – from the MNA and especially from the FLN – ended up killing probably more Algerian than French people."[13] As one Messaliste militant put it, according to Mohammed Harbi, "What I went through was horrible: I spent the whole war fighting Algerians."[14] Harbi notes that this same militant was later assassinated by his former comrades, concluding that "Algeria has lost a major part of its vital forces from internecine fighting."[15] Harbi pointedly questions the real motives for these internal killings: "Political factors fail to account for the relentlessness of the killings. Who is responsible for this tragedy?"[16]

A silent war between brothers was raging over who would become the leader of a state founded on dispossession and theft. However, it is worth noting that, in the frenzy of battle, it wasn't so much power itself that was being sought but rather the ability to deprive someone else of it. This logic of dispossession is of utmost importance as it sheds light on what was at stake in these power struggles. Indeed, it wasn't so much a matter of *having* as of *taking from* the Other the mere presumption of "having," especially since this Other assumed a role similar to the "father" – in other words, a provider, founder, or leader. Since the colonial rupture in Algeria, the father's disappearance has been endlessly re-enacted.

This explains why, throughout the history of political power in Algeria since the War of Liberation, each time a man is put in the position of father of this burgeoning nation, he is executed or removed from historical accounts. A paradoxical pattern has emerged: fighters keep calling for a unifying leader and yet, once the father figure appears, he is destined to be executed, repeatedly leaving brothers as either orphans or murderers. A turning point occurred during the war when the death and removal of the founding father (Messali Hadj) transformed into the murder of brothers. This is what history has clearly shown.

Suspicions about "who is who?" proliferated at the time.[17] Over the course of 1954, the population was for the most part in the dark about who launched the War of Liberation: the Messalistes or other militants? The Messalistes did little to dispel this uncertainty. This bolstered their own narrative about being the sole national movement. The FLN, however, went to great lengths to control the narrative and recruit fighters in order to be the sole power, targeting all real and perceived rivals. At its origin, this movement didn't have the financial, human, and material means to take on a powerful French army. The FLN ran an aggressive campaign both inside and outside the country to elicit support for the "national" fight and to position itself as the lone party leading the war.

In 1956, Messali Hadj narrowly escaped assassination and, in December 1957, Ramdane Abane, leader of the FLN, was assassinated in Morocco by his revolutionary brothers-in-arms. After this, the FLN's internal politics was guided by fratricide. In June 1957, a proposal for reconciliation between the FLN and the MNA was put together, explicitly calling for the end of "fratricidal" fighting: "Fighters and militants of both organizations are called upon to put an end to fratricide, which only benefits the enemy. ... Fighters and militants of both organizations are being asked to come together as brothers in order to join forces and hasten the liberation of the fatherland."[18] This text remained as a draft, and the fratricidal fighting continued.

In 1959, the President of Tunisia, Habib Bourguiba, wrote a letter to Messali Hadj pleading him to take on the problem of fratricide for the sake of the national cause. He acknowledged his place as the father of Algerian nationalism and positioned the revolutionary brothers as his heir-apparent sons:

> History will remember you as the father of Algerian nationalism. And in spite of all the repression you have faced, your work has created thousands of tried and true militants. Whereas the FLN's ranks today are filled with militants formed under the harsh training of the *Étoile Nord-Africaine*, before passing to the PPA, then to the MTLD. ... I have always advised you to forget (if just for a moment) your old grievances, your old disputes. ...You did no such thing. As

> a result, we now have this appalling spectacle of fratricidal fighting, of the settling of scores between compatriots.[19]

In spite of the various calls to join ranks behind the "national" *cause*, the fighting continued and progressively consumed the united cause from within. A sort of incurable disease was eating away at the "national" body. Its symptoms consisted of suspecting and hunting down perceived enemies and obsessing over conspiracies and supposed betrayals. This was inevitably followed by the physical removal of the enemy. With the high number of losses within the MNA, the internal fighting between the FLN and MNA slowed, but the war then spread among the ranks of the FLN itself. War steadily grew from within. The killing of brothers-in-arms accelerated at an alarming pace after the assassination of Ramdane Abane.

This assassination inaugurated the reign of fratricide. In other words, it marked the "rejection" of fraternity. During the FLN's first few years, Ramdane Abane was a unifying force and party leader. He put forth a vision for building an Algerian nation that would be plural and ruled by its citizens. His execution occurred right when he was recognized by the majority of FLN members as the new "father" of the Front. Those suspected of killing him include Lakhdar Ben Tobbal, Abdelhafid Boussouf, and Krim Belkacem, who would also be executed by his comrades, in 1970.

Detailing all the atrocities and internal attacks occurring within the FLN would require its own separate work. For our purposes here, it is worth citing the account provided by Ferhat Abbas, President of the Provisionary Government of the Algerian Republic (*Président du Gouvernement Provisoire de la République Algérienne*, or GPRA) and an important nationalist militant who dedicated his memories, *Autopsie d'une guerre*,[20] to Ramdane Abane. Abbas recounts how he himself was many times targeted and how his nephew was killed by the FLN in this period of fratricidal violence by being mistaken for him. He paints a chilling portrait of the internal violence raging among nationalist militants and is struck by this re-enactment of colonial violence, which draws on its same methods and techniques:

> The abuses we are guilty of are stains on the history of the FLN. ... The El Halia massacre, the red night of La Soummam, the executions in Melouza, the pointless killings and torture all could have been avoided. ... In countless cases, the behavior of some leaders and some resistance fighters was appalling. Innocent people were

> assassinated in order to settle old scores, which had nothing to do with the fight for Independence. We condemned France's use of torture, but we practiced it on our own brothers.[21]

The killings and executions among brothers wouldn't stop with the Independence. The toppling of power, assassinations, *coups d'état*, and imprisonment without trial continued in an almost identical manner.[22]

On June 19, 1965, Colonel Houari Boumediene, former general of the ALN during the war, seized power following a *coup d'état* (it is worth noting that he kept his *nom de guerre* while in power).[23] This episode, along with his decision to imprison the former President, Ben Bella, his ally up until the end of the war, marks the beginning of the reign of terror. On December 14, 1967, President Boumediene narrowly escaped being overthrown by his army's general, Colonel Tahar Zbiri. And on April 25, 1968, he was assassinated. It should be noted that Boumediene was often described as obsessed with being overthrown by a *coup d'état*. And with good reason. This was precisely how he rose to power. This history became a regular pattern. In 1992, President Mohammed Boudiaf was assassinated after rising to power following the overthrow of the government of President Chadli Bendjedid, who himself had been put into power by those who would later take him down. Nothing is known about those behind the *coups d'état* and political assassinations. Not a single one of them has ever been tried. History continues to be ruled by dismissal. Accounts are riddled with blank space, which governs the decisions and actions being made to this day.

The absence of an elite class or a leader in Algeria today is a direct result of this logic of removal. It is a matter of maintaining at all costs the illusion of a horizontal line of power. Leaders remain in the shadows, their power forever dependent on the removal of another, even if this Other belongs to the same group. This logic affects every level of society, from the social bond and the organization of institutions to the distribution and dissemination of knowledge. Power is aimed at dispossessing the Other, not so much in order to become the exclusive authority in spite of its apparent aim, but to ensure that the experience of elimination persists. A precise and well-regulated system of exercising one's self-proclaimed right governs this process, which depends on three operations: dispossession, removal, and then enforced deprivation.

Everyone at some point pays the price of this logic. Contrary to what many may believe, politicians and leaders are not immune from this: they also fall victim to the systematic seizure and the toppling of power.

When the murders between brothers are dismissed ...

Precluding alterity in the private sphere transforms the community of brothers into its own double. This explains the climate of suspicion in which various combatants (*djounoud*/brothers/*mujahedeens*)[24] lived during the war and its persistence after Independence. Countless executions were carried out on the basis of these suspicions after Independence by a secret service that diligently hunted down and executed suspects. Paranoia reached a new high during the first few years of Independence, with everyone becoming a potential enemy who threatened national unity. For this reason, each and every enemy had to be removed.

An Arab-Islamism (LRP) aimed to bridge these internal divisions by offering a unifying national and international vision. This vision was designed to pacify internal threats. But it did this by creating a vast and rigid foundation right over the gaping hole left by the colonial rupture. And the figure of the persecutor steadily gained in importance. This figure became present in each and everyone alongside its counterpart, the persecuted. Which means the subject was being hunted down and policed from within by a system of self-surveillance. The subject is led to identify with an archaic superego that leads by destroying anything that disagrees with its particular moral position, including, in this case, its own self.

We are not far from the notion of sacrifice, especially considering that those killed during the war are deemed martyrs (*chahid*). In Algerian history, Ramdane Abane is considered a *chahid* of the Revolution despite being assassinated by his brothers-in-arms. Note that this martyr has three tombs, either to ensure his true death or to protect himself from his own ghost.[25] Going from the target of assassination to a martyr of colonization is sadly not uncommon. Sacrificing one's own brothers in the name of an obscure god raises questions about the killing craze sparked by the disappearance of fathers under colonialism. The killing of one's own thus results from a profound absence of recognition and a legitimate existence.

Brotherhood gives way to heightened suspicion whereby one's fellow persecutor becomes an ally to be executed.

Even today, power remains in the hands of these fratricidal brothers. Whether they are alive or not is unknown, since the number of assassinations is shrouded in secrecy by the state. In this context, seeking power is a dangerous game played by fatherless brothers, who, as Jean El Mouhoub Amrouche says, are cut off from their "past and future lineage." This clan of brothers runs the government, shares the profits made from oil, and continues to adhere to a logic of removal. The competition to be the sole leader of the nation seems to be one of the major effects of the disappearance of fathers under colonialism (see chapter 2 above). Brothers are caught in an inescapable bind: on the one hand, they are doomed to remain brothers, as the position of father, destroyed under colonialism, remains forever out of reach; on the other hand, fratricide, nevertheless fueled by a search for the father, amputates the genealogical line so that brothers will remain nameless sons of disappeared fathers. Coloniality, as previously shown, hollowed out a paternal function that organized the social bond in traditional societies. Reducing fathers to worthless objects left abandoned sons with a caricature of the father in their leaders. Sons seek a father and, at the same time, ensure that he remains nowhere to be found, unwittingly re-enacting colonialism's practice of disappearance. Thus, they are seemingly condemned to exercise an illegitimate power.

Since the War of Liberation, the political being of each individual has been undermined by a profound feeling of illegitimacy among brothers. Forever cut off from their lineages, the above-mentioned fathers of the Algerian Revolution transformed into a band of self-destructing, blood-hungry brothers. The quest for absolute power and the cult of the hero spring from this feeling of illegitimacy deeply embedded in the subjectivities of sons. In this context, suppressing political plurality becomes a structural necessity to fill the immense gap left by an absent father. This absence of paternal function within a patriarchal system has major consequences for the organization of power, beginning with the collapse of the symbolic order that mediates between those ascending to power and those they claim to represent. Ferhat Abbas's writing clearly shows this shift to a state of turmoil felt by orphaned, fatherless sons in the face of power. This is especially the case given that the young Independent nation was founded on atrocities and abuses [*détournements*] committed in the name of power, which only reinforced the experience of illegitimacy. Lacking fathers, the brothers' struggle became a form of governance. As Ferhat Abbas has explained, "Algeria was like a wounded animal surrounded by a pack of wolves sharpening their fangs. Everyone wanted a piece. And too bad if in the end the country suffered as a result. The absence of a strong and legally recognized authority gave rise to a multitude of devious ambitions."[26]

The very particular structure of political power in post-Independence Algeria brings to light unusual factors whose interpretive consequences extend well beyond the country's borders. Indeed, this situation shows how all political power relies entirely on someone seen as performing the paternal function, who stands as *both an authority and a mediator*. This seems to suggest that all democratic organizations are bound to the two dimensions of authority and mediator, which themselves are a product of the paternal function. Lacking this, the political order becomes a *totalizing* regime that depends on theft. In this latter case, the leader, an obscene caricature of a father, nevertheless plays that role, masking its real absence. Unable to reconcile "father and nation," the very meaning of unity is called into question.

Mohammed Harbi points out that some fratricidal killings were a means for the FLN to reclaim unity. But a unity founded on murder, which remains precluded from most accounts, is fragile and under constant threat. Since the 2000s, there has been a clear effort to write this side of history, which has been erased by the political order. One could cite, for example, a recent work that examines the assassinations of Ramdane Abane and Mohamed Khider.[27] To this could also be added the memoirs of men who played a critical role in Algeria's higher ranks of power.[28] Speaking of nationalist combatants, Mohammed Harbi writes: "All they have right now to unify them is the clearly stated desire to be the *leaders of the Algerian state in the making*."[29] The stakes of these killings are clearly revealed: to be the sole and rightful leader of the Algerian state that one can claim as one's own. This search for ownership, however, is an illusion that hides a much more serious crisis: the never-ending destruction, undoing, and dispossession of the Other, of History, languages, and so on. Despite acquiring this power (of "having"), no one seems to feel politically legitimate in this position. The pseudo-legitimacy earned from having fought "heroically" for liberation always stands in place of any real political legitimacy.

The issue of legitimacy plaguing subjective and political lineages has only continued to gnaw a deeper and deeper hole. This deeply ingrained state of illegitimacy can only lead to crime, which in turn reinforces the feeling of illegitimacy. Nothing seems capable of stopping this vicious circle of sons who cast themselves as heroes, and, even more frequently, as state leaders. These roles fail to reanimate the paternal function. What's more, in the search to fill this coveted father role, the abyss only widens.

Calling on the father

Three overlapping aims shed light on the underlying stakes of the War of Liberation: becoming a leader (attaining the status of heroic son), rapidly expanding ownership, and seeking legitimacy. In light of this, it is worth revisiting *Totem and Taboo*, which functions on many levels by exposing one's relation to power, to desire, and finally to the political order, which structures the social bond. For Freud, as we will see, killing the father who is in the position of tyrannical leader forever bars the brothers from murder and ownership. But the political (and subjective) situation in Algeria counters the very premise of this myth. In contrast to Freud's account, killing the leader – in the "national" narrative of Algeria – neither prohibits killing nor curbs the brothers' desire for ownership. Far from it. The father, as a mediator and provider, cannot be found. The founding fathers of the "nation" – Messali Hadj and Ramdane Abane – were taken out, which drove the brothers of the Revolution to kill each other. Is this the hidden side of the social order? And what happens when an anarchic and destructive death drive is the bedrock of the social bond?

In Algeria, the bloodthirsty side of the "brothers of the Revolution" did, in fact, become the dominant force, disabling political action and the possibility of social harmony. The law of fratricide became widespread and translated into murdering almost anybody, giving rise to a post-Independence civil war fueled by the indiscriminate killings of "Europeans" and Algerians, before becoming a form of self-annihilation during the Internal War. The reign of fratricide, it could be argued, depends on a military regime to prop it up and, in so doing, also invalidates the function of the state. Let's recall that in Algeria the FLN continued to tear itself apart in order to ensure the military maintained power over the political order. This is what led to the assassination of Ramdane Abane, who argued that the military should be reined in by the political order.

Not coincidentally, this law of fratricide is at play in the rise of totalitarianisms throughout the world. The two are so intricately bound together that one can no longer tell if fratricide gives rise to totalitarianism or vice versa. And so what happens to the social bond under the reign of fratricide?

Understanding the larger context of the War of Liberation shows how fratricide seems to target the site of the missing father. The murder and disappearance of fathers under colonialism has resurfaced with the killing – symbolic or otherwise – of the founding fathers. Tied to colonialism in this way, the brothers of the Revolution don't see themselves as guilty of murder: they attribute responsibility to the Other of colonialism. Which isn't entirely off base. Paradoxically, executing the founding fathers is a way of appealing to the father for the sons/brothers of the Revolution. It is an appeal made via implication, appearing as the act's negative.

Let's imagine for a minute how this is played out: these orphan and "bastard" sons (Jean El Mouhoub Amrouche) seeking legitimacy turn toward the father, and find an abyss that draws them in and instigates their own homicide. The sacrificial offering is made to a father who exists as a sort of phantom limb which, although absent, cries out in the darkness. The question "Who is the father?" (Kateb Yacine) is immediately followed by another one: "Where did the fathers go?" And why aren't their assassinations by their own people recorded as acts of murder? The brothers of the Revolution persist in seeking a father, but find only absence. They go to war over the killing of fathers under colonialism, but forget that they are simply re-enacting the same drama as they seek to become the sole leader with absolute power.

In *Totem and Taboo*, Freud explores this myth that explains the birth of society by the killing of the leader. He shows how subjectivity and the political order are both founded on this murder. This coincides with the invention of human civilization, which, for its part, depends on renouncing murder. Repressing murder and displacing it to the realm of fantasy allows humans to live in social harmony. Incest and killing remain forceful desires in the psyche insofar as they are prohibited in practice. As a consequence, according to Freud, parricide should logically lead to a taboo on fratricide. The guilt of sons is supposed to counteract their fratricidal desire.

We could say that once upon a time there was a jealous and tyrannical leader who possessed all the women of his tribe. Only he had unlimited access to goods, women, and power. This leader used his power to target his own sons, dispossessing them of everything and turning them into spectators of his unbridled pleasure. Overcome with jealousy, the sons dreamed only of one thing: to kill their father and finally free themselves from this tyrant who held them in a state of destitution. It is in this context, Freud explains, that the "expelled brothers joined forces, slew and ate the father, thus put an end to the father horde. Together they dared and accomplished what would have remained impossible for them singly."[30] The sons, once the act of murder has been accomplished, commemorate it by prohibiting themselves from what they wished to do (enjoy their various possessions), which the leader had also deemed taboo. Freud notes that the paternal function appears in this commemoration – an introjection of guilt – through the trace of a memory that curbs the sons' murderous desires. It is at this point that the community of brothers is founded on the taboo of murder and incest.

Instead of taking the place of the father, the sons pay homage to him by continuing to observe the taboos he had established. As with desire for incest, the desire to kill becomes part of psychic and social life. Desire is fueled by taboo, which makes its realization impossible. Except that this process doesn't happen once and for all. The subject is constantly agreeing to observe these taboos so as to continue feeling driven by desire. There is therefore a precarious relationship with the law, which is at the root of the social order. Without this law, barbaric behavior would be inhuman, whereas, in reality, it is what defines humans. Religion, morality, and culture teach and enforce these taboos so as to maintain this relationship to the law.

Fratricide implies a disabled state of the symbolic law while appealing to the father to set limits. In the context of Algerian politics, it could be argued that the absence of the father and the fact that he is missing (rather than dead, which can be commemorated) plunge sons into a deep state of despair where there are no internal limits.

In Freud's account, everything conspires to push the jealous sons to kill each other in order to occupy the coveted position of leader. And yet that doesn't happen because the memory of the defunct father serves to check their desire. Freud rightly notes that this idea is counter-intuitive. It seems only natural that sons freed from their father's tyrannical hold would indulge all their desires. There is thus an *apparent* contradiction in Freud's text. However, in a footnote, Freud alludes to another argument that undermines his own premise. This argument also helps explain the issue of fratricide in Algeria. He cites the work of the ethnologist James Jasper Atkinson, who noted that, in some cases, the sons' killing of their father can lead to fratricide, which is to say, to the opposite of the taboo on killing. After killing the father, the sons will continue to kill each other and nothing will stop the ensuing violence. The political situation in Algeria is closer to this account.

"Atkinson, who spent his life in New Caledonia and had unusual opportunities to study the natives," writes Freud,

> also refers to the fact that the conditions of the primal horde which Darwin assumes can easily be observed among herds of wild cattle and horses and regularly lead to the killing of the father animal. *He then assumes further that a disintegration of the horde took place after the removal of the father through embittered fighting among the victorious sons, which thus precluded the origin of a new organization of society.*

Atkinson, cited by Freud, concludes that this would lead to "*an ever recurring violent succession to the solitary paternal tyrant by sons, whose parricidal hands were so soon again clenched in fratricidal strife.*" Freud appears at pains to reconcile this with his own theory: "So much for the very remarkable theory of Atkinson, its essential correspondence with the theory there expounded, and its point of departure which makes it necessary to relinquish so much else."[31]

A gap in memory sets off an endless deadly battle

Freud's nuanced reading of Atkinson approaches the matter from another angle: there is no trace of memory leading the sons to the father, leaving them with no one to commemorate, honor, much less to love. *Neither memory nor love* ... Then the question becomes: how can a father be invented in order to maintain the taboo on killing? Atkinson's remarks show how fratricide can prevent the coming of the father. There is indeed an appeal made to the father, but he fails to materialize owing to the infinite re-enactment of murder. The paths of memory are obstructed and the traces of the father have been obscured. His memory cannot be found.

After the colonial rupture occasioned the disappearance of fathers, who were seen as mere objects, sons were thrust into an impossible situation where fratricide was viewed as a natural consequence. Colonialism, as shown, made the father disappear without a trace in order to honor his memory. Sons were deprived of the possibility of killing the tyrannical leader. The colonial Other took care of his disappearance and of erasing all traces, relieving the sons from the responsibility of committing this act. The murderer is the colonial Other, and the father was taken from them before any memories could be formed. There is a clear appeal made to the father, but the abyss caused by his disappearance inspires self-annihilation and above all the re-enactment of killing whomever aspires to the place of father. The appeal made to the father is met only with the echo of the brothers' voices.

In Algeria, seeking legitimacy by wielding power is an unmistakable appeal to the father. However, this legitimacy isn't conferred by one's knowledge of Modern Standard Arabic (deemed sacred in the Qur'an), or one's war experience, or even through one's religious zeal. The killing and disappearance of the father remain unacknowledged, and, for this reason, impossible to commemorate and therefore forget. Fratricide, like the father's disappearance, has been granted a dismissal. Fratricide is the bloody wound at the site of the unburied father's disappearance.

Kateb Yacine made his heroine, Nedjma, the child of adultery and crime, who unwittingly engages in incest with her brother. In the novel, she is a metaphor for a dispossessed, embattled Algeria. This explains her death, and also her victory, since "she triumphs in all her beauty."[32] Let's recall that the plot of the novel centers on the father. But who is Nedjma's father? Her father unknown, Nedjma is transformed into an ogress who is susceptible to crime and fratricide: "Nedjma the ogress of obscure blood, like that of the black man who killed Si Mokhtar, the ogress who died of hunger after eating her three brothers. ... Nedjma the trembling drop of water that swept Rachid off his rock, drawing him toward the sea."[33]

From this gaping hole left by an ogress who kills her brothers emerges madness, a sort of mix between a delirious hallucination and "a dream beyond memory."[34] Kateb Yacine and Nabile Farès both document in their writing the site of the father's fall from memory. In contrast to the Freudian myth that posits a near equivalence between father and memory, colonialism effects the disappearance of the father and the erasure of memory. Therein lies the real crime of colonialism, thrusting the sons into an endless deadly fight. As a result, a gaping absence remains and a cry gets lost in an infinite, memory-less vacuum, leaving no legible traces whatsoever. "Come and take a look and see the shape-shifting vacuum of being"[35] (Nabile Farès). Since the father's fall and erasure, sons are "untethered, tribe-less, in the vertiginous space of a lightless night, beyond the stars, with the only baggage being an *absolute lack of memory.*"[36]

The colonial practice of erasure makes it impossible to discover the traces of what took place. The historical record that would recount the disappearance of fathers is scrambled. This situation creates an enormous gap in history and memory around which sons fight each other *as brothers*. With the father's disappearance, Nedjma, the unmarriable woman, takes on the role of mother, as she is forbidden from masculine desire. Everyone wants to possess her. She leads to the path of incest and sets off a frantic race among men ready to pounce on her and possess her. Each aspires to the coveted position of rightful owner, sole possessor of all privileges and leader by way of theft. When the paternal function fails to materialize, fratricidal murder becomes an incontestable reality, impossible to forget and impacting the social order. It is the trace of the father's disappearance.

In Algeria, the phenomenon of fratricide has a forceful impact on political and psychic reality. Failing to recognize this impact within the clan of brothers and its ties to the disappearance of the father, it remains active as the key organizing drive toward social harmony. Is the fratricidal tendency underlying political power therefore the only memory of colonial trauma? Maybe more than three generations – three plus one? – are needed to restore the father's name, give him back a body and then a burial so that a Republic marking its departure from the bloodthirsty masses can be born and flourish. And to hasten this *work in progress*[37] let's start by admitting, as Kateb Yacine has said, that "*l'enterr'ment di firiti I la cause di calamiti*" ("the burial of truth is the cause of calamity").[38]

Fratricide in Algeria acts like the realization of a memory from which remembrance is excluded. But, even outside of Algeria, contemporary history has shown that all revolutions are born from the unity of brothers in the face of a tyrant. In other words, unity is formed out of an opposition from which it becomes impossible to escape, making it hard simply to be *with* the Other. In Algeria, discourse is marked by this embattled state of being *against* … Arabism, Berberism, French assimilation, and so on. Which is to say, against the Other, and, by extension, against one's self.

6
The Internal War of the 1990s

In consequence of this primary hostility of human beings, civilized society is perpetually threatened with disintegration. ... Instinctual passions are stronger than reasonable interests. Civilization has to use its utmost efforts in order to set limits to man's aggressive instincts and to hold the manifestations of men in check by psychical reaction-formations.

Sigmund Freud, 1929[1]

The city discarded its memory and now,
faced with the agonizing void,
it is looking for a new memory,
borrowed from other cities,
near or far. ... We need an unbounded, undaunted love to express reality.

Waciny Laredj, 1993[2]

On March 8, 1938, Arabic was declared by decree a "foreign language" within the colony. Algerian writers, as we've seen, flouted [*détourné*] the political organization of languages (Arabic, French, and Tamazight) by smuggling within French banned and neglected languages. This was a powerful and extraordinary move that mocked the censors by upsetting the boundaries of the familiar and the foreign. Henceforth, it was determined, never again would a language obey a political definition in service of exclusion and repression. This repudiation was the inaugural moment of Algerian literature, which casts itself as a reservoir of living memory. The rejection of subjugation is still felt today among contemporary Algerian writers (who write in both French and Arabic). This initial gesture reminds us that no language is inherently oppressive or oppressed.

Reconsidering the LRP bloc

Any language has the ability to mend and heal the worst human wounds. Those in power were aware of this in the immediate aftermath of Independence. There was a sense of urgency to rediscover languages and of hope that they could restore the shattered interiorities of the living. And so, taking heed of this, the political decree was reversed. Now it sanctified language, deeming it powerful and untouchable, in direct proportion to the way private languages had been beaten and raided, going so far as to place language out of reach. It didn't take long for a beneficial idea to work against its own initial insight. The function of languages was radically overturned. Modern Standard Arabic was declared the national language. This decision offered many historical and political advantages.

This language conveyed a feeling of unity and a sense of belonging to an (Arab) world; it demonstrated an inclusiveness that countered the exclusion practiced by colonialism. Indeed, deemed the language of God – one that was uniquely capable of revealing the message of the Qur'an[3] – it was bestowed a power within the imaginary that made the colonized subject feel less vulnerable. Reduced to this false religious understanding, this language nevertheless offered the illusion that it could protect subjects from helplessness in the face of the foreigner/enemy.

Held in the highest esteem by the political order, it created potential limits by that very fact to the intellectual freedom of subjects at the time of Independence. The primacy attributed to Arabic – which relied on confusing language and religion – caused mother tongues to be relegated (once again) to the status of "dialects," deemed contemptible and unworthy for thought and knowledge. In this way, thinking also came under political control. Thus, the language of the master, French, was replaced by a language with an uncontestable power: that of God. This explains how opposition to French colonization only gave rise to colonization of a different sort (Arab).

The use of Arabic benefited the political order by wiping out the past, especially the linguistic hybridity that once characterized all of Algeria. It eradicated in this way the many points of contact between languages and religions (Algerian Arabic, for example, with its mix of Tamazight, Spanish, Judeo-Spanish, Turkish, and French). The only solution to curbing the anxiety provoked by freedom regained (or simply gained) was a return to Arab-Islamic colonization.

The questions raised by Algerian literature since colonization – "Who am I?" and "What am I?" – were met with a definitive political response that left no room for doubt. A profession of faith dating back to the War of Liberation has been imbued with an irrefutable certainty: "Algeria is my country, Islam my religion, and Arabic my language." This message alone shows how Independence distorted [*détourné*] Algerian history to meet its own needs, relying once again on blank space and subjugation. The ancestral past of its myths, embedded in, and transmitted by, other languages, especially Tamazight, was dismissed. Power, languages, and religion were all hijacked [*détourné*]: the first by a series of *coups d'état*, the second by the political function of Modern Standard Arabic, and the third by the political order. As explored in the first chapter of this book, the LRP (language, religion, and politics) bloc was formed by a masterful use of *détournement*. This unfolded at a time when a very strong nationalist ideology was drowning out the fundamental work of archiving and creating a pluralist history. My argument is that this ideology relied on erasure and homogenization to mask the practice of fratricide that was ravaging the country. Indeed, executions were being carried out among members of the same clan (brothers of the Revolution). Several decades later, this internal war over power still rages on and is reflected in the frequent dismissals of generals and other army commanders. The military regime ensures that fratricide and deal-making between families remain in full force.

Imposing Arabic as a sacred language poses many problems for treating the wounds of "colonized" subjects. For one, it prevents mother tongues from flourishing and playing an active role in the culture. Contempt for languages that define subjects has returned under another guise. Placing Modern Standard Arabic at the top of a linguistic hierarchy was designed to encourage its occupation of "mentalities." This linguistic war signaled an awful devastation to come. From the vantage point of the present, the imposition of Arabic appears as a vindictive response to colonialism. It was a refusal of an education taught by Algerian poets and writers *in French*. Language, religion, and politics (history) should thus be nationalized. And so the LRP was formed. Removing languages that bore witness to another history – Algerian Arabic and Tamazight – allowed for the creation of an origin myth that concerned only two warring languages: Modern Standard Arabic and French. Other languages were left out of this story.

Building the LRP around Modern Standard Arabic was a political move that made French colonization ground zero in the land's history, as though nothing had existed before the Arab colonization. The same kind of colonial logic resurfaced that sees conquests as occupying land that has no previous history, knowledge, and culture. Liberation was thus followed by a linguistic war and a *totalizing agenda*. Little by little, the process of Arabization, driven by a desire to make French disappear, created cleavages between languages and generations. This has had a major impact on the social order. For example, once again a barrier was erected between parents who used French and children who thought in Arabic. Hence the inversion of language use between generations, not unlike what happened during colonization. Indeed, children who had been educated under colonialism read and wrote in French whereas their parents were confined to languages deemed unworthy of use, if not outright banned. This explains the struggle and shame felt by children in regard to language, as *hogra* became an inevitable part of their linguistic experience. Speaking the language of their parents, they were part of an "*indigène*," or even undignified, family line; and, speaking the language of the oppressor, part of the master's lineage. In both cases, language conveyed war, shame, and *hogra*.

If there is "totalitarianism" in Algeria, it could be argued that it developed by turning *language, history, and religion into a single totality*. This is what is meant by the notion of an LRP bloc, which works to suppress social harmony, political life, and psychic space by destroying the conditions that make them possible. But the everyday speech of Algerians, made up of many languages – Arabic, Tamazight, French – demonstrates the resistance with which this is met. In the majority of the population, the speaking subject introduces a plurality in place of a totalizing agenda, which, for its part, shifts to another arena, lending itself to the realm of political censorship. This everyday speech joins Francophone Algerian literature in making language's alterity tangible, bringing it to life right where political power (colonial and post-Independence) repeatedly seeks to exclude it. In this landscape, speech and writing push toward fragmentation in the face of a totalizing agenda.

The tyranny and pleasure of power

Frequently, a leader's fall from power reveals in a display of violence an underlying fratricidal struggle. A moving novel by Rachid Mimouni, *Une peine à vivre* (1991), illustrates this essential aspect of power.[4] It tells the story of a near-illiterate man who, by dint of guile and resourcefulness, gradually climbs up the military hierarchy during the War of Liberation to become a leader – a "Grand Marshal," to be precise. He reaches that position after assassinating the previous leader. This character is obsessed with assassination plots, while he himself is skilled at orchestrating them. The leader constantly gets rid of any person who thwarts his unbridled desire for possession. He does this by subjecting them to "the polygon [i.e. firing squad]." Under his reign, murder is law and the country is governed by the leader's whims. But then the tide turns: the "Grand Marshal," despite his obsession with assassination plots, finds himself forced to "face the polygon" by one of his subjects, who was a comrade-in-arms during the War of Liberation.

Mimouni shows how totalitarian systems persist by cultivating a pleasure in fratricide. The leader's death offers the possibility of playing out this fantasy. Once the leader is assassinated by his closest allies, fratricidal murder manifests itself again. And for good reason. All leaders acquired their place illegitimately, through assassination plots or by overturning power or through killings between warring comrades. And so the cycle continues.

Overcome by his terrifying and tyrannical power and in love with a woman, the "Grand Marshal" decides to withdraw from leadership and organize democratic elections. With this, a serious danger threatens the system. His comrades in power reject his proposal, pointing out that he's risking his life and undermining his own people in the regime, whereas these same people are dreaming of taking his place. He is left without many options, as he is convinced that sooner or later he'll be made to "face the polygon" by his "plotting" comrades, or that, simply by retreating from his position, he'll still be assassinated by them. All this to keep the killings by those in the military cloaked in silence. The man who sowed terror, who saw himself beyond the reach of death, has no other choice but to face one death or another: "I also know that the terror I provoke in you," the Grand Marshal will say,

> will forever trouble your sleep and rob you of the desire to live long after I'm gone. You may see my limp body collapse on the ground, come and march before my corpse, spit one by one on my purple face, stare at my wound, feel my stiff body, but you won't believe I'm dead. You may bury me as deep as possible in the earth, pour infinite layers of concrete on me, incinerate me, and scatter my ashes to the four corners of the globe, or chop me to pieces, but I'll come back to haunt your nights.[5]

The leader cannot be killed; his power is unlimited. This isn't specific to a particular leader, but to the leader function. What matters simply is to keep the system in place, repeatedly cycling through illegitimate leaders, as each reaches his position by killing rival "brothers." The system's omnipotence is felt by all, defying time and death. This book was published in 1991 just as the Algerian political order was trying to escape a system of governance very similar to the one described in the novel. Thus there is some truth to this story of the hunted leader that reveals itself in the form of a projection. In other words, the leader attributes his own schemes to others. Since Independence, the political order has repeatedly been formed by overthrowing those in power and by making deals between families. This has only further stoked conspiracy theories. Having fought in the war – both the War of Liberation against France and the Internal War fought over the possession of "Algeria" – served as a justification for the seizure of power. This move, as we've seen, worked by erasing the role fratricide played for these "heroes."

What role did Islam play in this hero-and-martyr logic? Did it help or hurt fratricide? At this point suffice it to say that Modern Standard Arabic and Islam were signifiers of the political regime that were used to claim an otherwise inaccessible legitimacy. At the same time, in the background of all this, these signifiers colluded to erase history, languages, and plurality in favor of provoking subjugation to the highly sought-after ONE: One God, power, language.[6]

In *Une peine à vivre*, the tyrannical leader is taken hostage by his own power. His falling in love with a woman rattles the entire system of power. This bolsters the argument that the feminine figure stands in the way of fratricide. Removing this feminine dimension thus serves the pursuit of a specific pleasure: the *pleasure of fratricide*. Right when the "Grand Marshal" decides to give up his power, he has a surprising conversation with one of his subordinates:

> "If you leave, what will happen?" his subordinate asks rhetorically. "The country will be plunged in chaos."
> "Don't worry, we'll make it a smooth transition."
> "But there is no one to take your place."
> "With your fawning behavior, I'll end up believing I'm the Messiah. In spite of where I stand now, I haven't forgotten where I came from. And I still believe that any man randomly chosen could be a Grand Marshal who is just as good as or better than I am. … Can you tell me what I have achieved that's so remarkable since coming into power? I simply followed the lessons of my predecessors by ruling with absolute terror. I spent all my time taking down or assassinating anyone who threatened or displeased me. I governed according to my whims, which is to say, without remorse or mercy."
> "If you abandon the palace, the country will suffer a civil war."
> "Why?"
> "Those competing for your place will come forward and waste no time in tearing each other apart. Military leaders will march up to the palace with their troops behind them. There will be an explosion of plots, schemes, betrayals, unlikely alliances, new coalitions formed overnight."
> "I'll make sure the handover goes smoothly. … People will be entirely free to choose my successor in open elections. Backed by the people's vote, he'll have nothing to fear from me."
> "You've gone mad, Grand Marshal. You haven't weighed all the consequences of your decision. You think that the military will let any civilian with a troubled past take over the palace and give them orders after a simple ballot count? We wouldn't even be able to call him Grand Marshal."[7]

This passage shows how difficult it is overturn a system built on the pleasure of fratricide. Despite the leader's good intentions to set up a democratic form of governance, some things remain impossible. The core of the system is untouchable. Mimouni raises a very pertinent question: how can you escape fratricide when it has becoming the reigning law? This wonderful book sheds light on a unique moment in Algerian political life.

The shift of 1988 and the experience of political plurality

On October 5, 1988, serious protests broke out in several cities. Disaffected youth, high school and middle school students, workers from every social class and profession, women and men of all ages, took to the streets. They were demanding the end of a single-party government (the FLN), freedom of speech and thought, and the end of systemic oppression. These events took place amid a social and economic crisis. Demonstrators were clamoring for the need to escape a power structure and social organization driven by *hogra*. The general public wanted to break with a form of internal colonization and systemic servitude. Demonstrators harked back to the War of Liberation and sought to hold those in power accountable.

These protests soon turned into riots. Young demonstrators attacked state institutions, burned cars, and raided symbols of wealth that represented, in their eyes, an oppressive and unequal society starkly divided along social, economic, and even linguistic lines. Those in power – most frequently Francophone – addressed the population in a foreign language: Modern Standard Arabic. They were therefore using a language that served as an obstacle to dialogue. The general public called for re-establishing bridges between politics, history, and living languages, and demanded an end to a political power driven by exclusion, *hogra*, and abuse of power. This caused the military to fan out across Algiers and shoot into crowds, leading to many deaths (150 according to official accounts and 500 according to independent associations). Torture was also used on a large scale on both the young and old who had made an entirely legitimate demand: to improve living conditions and end the system of self-annihilation.

Two days later, on October 7, a wave of Islamists invaded the streets of the capital. These protesters were chanting, "For Her we die, for Her we live" ("*Alayhia nahia wa elhia na mout*"). "Her," in this case, referred to an Islamic state that had become personified. The word *dawla* (state) is feminine in Arabic. This gives another interpretation to power's desire for possession. It was a matter of fully and completely "having" her (the state). Another slogan was aimed at the FLN: "Down with the party of France" (*Berra hizb França*) – the party that had actually launched the war on November 1, 1954, and defeated France. An embattled political order translates into embattled languages. A frightening reversal of History was taking place, and some pro-democracy leaders, mainly Francophone, found themselves placed on the same level as members of the FLN government.

The Islamists' protests through the streets of Algiers were spectacular to behold. Men falling over in a trance implored God in tears. Thousands of people were praying in the streets. Those who had protested two days earlier demanding a democratic government were seized by fear and worry. Many could be heard asking: "Where did these Islamists come from? Where were they before?" For years, Islamists were working in the shadows and showing up when the state failed, making up for the gaps in the social and health systems. They brought aid, money, and resources to a people in need. Many among them, including several university professors, had fought in Afghanistan against the Soviet Union in the 1980s. At the time, the government showed little concern for this support, and even expressed hope that this exodus of "rage" would relieve the spread of Islamism in the future, without giving any thought to what would happen when they returned, as the writer Mansour Kedidir has noted.[8]

History will show that this activity was driven by *jihad*. This makes it one of the first appearances of a jihadist network with an international reach. The younger generations of jihadists throughout the world belong to a lineage that dates back to this period. It is interesting to note that some Islamists in Algeria went by the name "Afghan." And a well-known mosque in the heart of Algiers, named "Kabul," long served as an important center of activity. Some Algerian Islamists were Bin Laden's traveling companions at the end of the 1970s.[9] These same Islamists first infiltrated France via an affiliated network and then spread throughout Europe: Great Britain, Germany, and Belgium.

In Algeria, the Islamists steadily rose to power by reaching out to vulnerable populations that had been spurned by the state. They stoked hate for the political establishment, whose failures hardly needed pointing out. These behind-the-scenes operations didn't pose any serious threats as long as political office wasn't being sought. A multiparty system then emerged and Algerian political life was undergoing a massive transformation in under a year (between 1988 and 1989). In 1989, a new constitution was adopted and in 1990 the first multiparty elections were held. In December 1991, the FIS (*Front Islamique du Salut*, or *Islamic Front of Salvation*) won the first round of legislative elections in a surprise victory (with 54.25% of the votes).[10] With these unexpected results, the government, which had allowed this party to form legally, decided to cancel the second round of elections. Pro-democracy advocates were divided into two camps: those who supported this political decision and found themselves against their will aligned with the political establishment they denounced; and those who, committed to the principles of democracy recently established in Algeria, supported pursuing the second round of elections, thereby taking the risk of precipitating the rise of an Islamic state. Pro-democracy advocates were caught in a real stranglehold as both sides (Statists and Islamists) strove to realize their own totalizing agenda. Of course, their totalizing practices were at odds, but both were effective at stifling and reversing any advance toward democracy.

It was in this context that the army forced President Chadli Bendjedid to step down. He was replaced by a college council. The president's departure shows how the birth of democracy was accompanied by a *coup d'état* led by the military regime. For many, the army's intervention was welcomed and even encouraged, as though they suddenly forgot that the initial goal was to escape military rule. The so-called major "choice" was between allowing the FIS to set up a totalitarian Islamic state or accepting the state's own drive toward totalitarianism. The Islamists took control of the LRP bloc, solidifying their power through their promotion of Sharia, or Islamic law.

The whole affair seemed scripted. Indeed, it was during the first free elections that those who were calling for them found themselves stuck with an impossible choice: accept the results of the democratic process, which would put in power a violent and totalitarian regime, or refuse it and find yourself allied with a regime you had denounced. It is interesting to note that, in both cases, the same trouble presents itself: no matter which path one takes, there seems to be no escaping totalitarianism. In Algeria, the beginning of democracy revealed at the outset its own limits. This situation raises many questions: why does Algerian politics repeatedly find itself caught in this position where there is no room for negotiation and no third way? Doesn't this division among pro-democracy advocates attest to a wider split that masks the fact that, for both parties (among those in favor of democracy), they are still obligated to support a totalizing system in spite of their effort to move away from it?

An internal war of unprecedented violence

The division between "democrats" (those in favor of democracy) offers a rich lesson. It shows how the divisions between "them" and "us" (colonizer/colonized, state/Islamists, democrats/state, democrats/Islamists) continue to shift and how, at the same time, the divisions never hold. Forever caught in the "same" position, the third way, which would offer a real possibility of an "independent" politics, has been abolished. The specter of totalitarianism consumed the so-called "democratic" sphere entirely. It was at this juncture that an internal war broke out with unprecedented violence.

The FIS insisted in the most unambiguous terms on the importance of belonging to an Arabo-Muslim, anti-Western, and fundamentally religious world, one whose language is closer to God's. This allowed them to accuse the establishment and the democrats of being "impious" and foreigners to God's language. In their version of history, the democrats and the men in power were the successors of the colonizers and the "*harkis*." For the Islamists, the Arab States were for the most part deemed "handmaids of the West."[11] At the center of the fight for power between the Islamists and the political order were competing ideas of how to proceed after the War of Liberation. For the Islamists, the Islamic state was the only valid solution to "cleanse and purify" minds of colonial contamination, which referred to Algerians who identified with French colonizers at the moment of Independence. The success of the Islamic political strategy lay in its being seen as a remedy for colonialism for the disaffected masses. They were the ones who embodied the *hogra* of the political order.

The Islamists thus drew on the signifiers of power to address the questions on everyone's mind: "Who am I?" and "What am I?" To this they responded – in complete agreement with their initial political goals – by offering a purified and original identity. The Islamists were transforming a totalizing agenda into a totalitarian system where the voice of God replaced that of the (military) leader. The same signifiers were at play (language, religion, and politics: LRP), but their instrumentalization for political purposes was pushed to the extreme. The same war was playing out, leading both democratic camps to come together, in the name of democracy, in order to create a totalizing power. A system emerged from this where two equally unfavorable entities were competing over the same signifiers. The novelist Arezki Mellal summed it up this way: "We're forced to choose between those who think the military should have ceded the country to fundamentalists, and those who think the military is a republican institution that saved the Republic."[12]

It was during this time of being caught between two sides of the same coin that Mohammed Boudiaf, who had fought in the War of Liberation and who was imprisoned and removed from power in its wake, was called upon by military leaders to become President of the Republic. During this crisis, the sole man who delivered his speeches in Algerian Arabic showed himself to be very concerned with opening up the political order to dialogue. He seemed the best fit for creating a third way, both through his positions on matters and through his willingness to bring the larger population into the discussion. He wouldn't, however, escape the cruel fate of fratricide. President Boudiaf was assassinated while delivering a speech in public on June 29, 1992, just as he started identifying the channels of corruption run by the "brothers" in power. He was killed by a member of the presidential guard, a second lieutenant in the secret service's special intervention group. (One year later, in 1993, the former head of the secret service was also assassinated.) Even today, much about the President's assassination remains unknown. Despite the evidence of fratricide, it was judged – behind closed doors and barring a civil justice ruling, like many other cases – an isolated act, which is to say, in this case a dismissal for the political order. The President's children later indicated that the military leaders and brothers-in-arms during the war had clearly ordered this murder.[13]

It was during this time of insecurity and confusion that the Islamists, kept out of the political game, took up arms and embarked on a path of resistance, plunging Algeria into a bloodbath. This situation, which afflicts many other parts of the world such as the Middle East, raises several pressing questions: how can we understand the structural relation between democracy and civil war? Do they go hand in hand? To what extent can Algeria offer a paradigmatic example for understanding the conditions and limits of democracy?

These questions are crucial to Europe, too, where the rise of extremism in democratically aligned societies poses a serious threat to the political order. Throughout the world, people are haunted by the specter of civil war. Jihadism is a symptom of a weakening democracy. In Algeria, with an Islamic state, social relations and legal questions would have been regulated by Sharia (Muslim law). The Islamists' political position was presented as the sole solution for healing the wounds of colonialism by restoring the glory of a Unique God who speaks through men. In the words of the Islamists, men are viewed not as interpreters of the sacred text but as the chosen ones who can reveal the word of God. Let's recall that the LRP used the religion and language of Arabs to thwart the emergence of a democratic state and secular thinking that would have undermined the band of brothers. On the other hand, political control was also hampering the Islamists, with a Minister of Religious Affairs whose mission was to control speech. In spite of the safeguards established under its totalizing control, the system was overwhelmed by what it itself had created and sought to tame.

The curse of fratricide

The Islamists knew how to turn this logic on its head and sought to replace the brothers in power with a single elected party (FIS) which brought together *Ikhouas* (Muslim brothers). During these years, the Islamists imposed their own system of self-reference: *khouya* (my brother), *ikh* (brother), or *okhti* (my sister), *el okht* (the sister). They sought to de-sexualize social relations. That only increased the likelihood of murder and incest, which rose dramatically during the Internal War. As early as 1978, Mahfoud Nahnah had founded the first Islamist movement in Algeria with the signifier "Brother" in its name. This underground movement was called "Brothers in God." Many (blood-related) brothers later joined the movement. This phenomenon isn't unique to Algeria. Indeed, the means of recruitment are popular throughout Europe: what role does fratricide play in so-called Islamic "extremism"?[14]

This divided brotherhood was clearly illustrated during the Internal War when, told that his son had died, a father responded: "Which one? The terrorist or the police officer?" Brotherhoods were thus fractured and killing each other, with one side working for the state (or simply finding themselves as citizens at the wrong time) and the other side "terrorists," sometimes unwittingly. This family drama perfectly mirrored the division afflicting the "democrats." Which of the two brothers is a "terrorist" and what kind of "terrorist" is he? Which of the two camps was more democratic: the one associated with the state, or the one associated with the Islamists?

This account has been told by many journalists and writers. In Rachid Mimouni's novel *La Malédiction* (1993), for example, two brothers are at the center of the story: Hocine and Kader.[15] Hocine disappears, and his family searches for him for several years. Kader, an obstetrician at the university hospital in Algiers (Mustapha Hospital), realizes overnight that he is working under the orders of the Islamists. The Islamists see to it that obstetricians never perform any deliveries for women out of wedlock. One of these women dies from this lack of care. The medical files are confiscated. Kader refuses to police morality and therefore ignores this law. Thus he steals the medical files in order to continue to provide care for patients and to keep their information confidential. But then he finds himself summoned by the Islamists, who waste no time in imprisoning him. To his great surprise, during sentencing, he comes face to face with Hocine, who is now a judge on the Islamic court.

Kader's case is thus to be judged by his brother. Their father, who fought in the War of Liberation, had been secretly killed in Morocco by his brothers-in-arms over a rivalry for power – which recalls the assassination of Ramdane Abane in Morocco as well as many others at the time. Fratricide is thus passed down to the next generation. But this time, it doesn't concern "brothers-in-arms" but real, blood-related brothers. Killings of this type routinely appeared in the papers during the Internal War. Frequently, these sons were the descendants of fathers who really disappeared during the War of Liberation. Their problems were inherited. It should be pointed out that the father's killers made this crime look as though it was carried out by the French army. Once again it is a question of a martyr (*chahid*), regardless of the conditions. The criminal, however, will always be the Other, hiding the reality of it being an internal affair. This confusion was effectively instrumentalized and assisted by the cult of martyrdom.

The war from 1992 to 2000 that followed the interruption of the elections was clearly an internal fratricidal conflict. In Rachid Mimouni's novel, the face-off between Hocine and Kader is both moving and shocking. On the one hand, this is due to the odd situation where a man is forced to judge and order the death of his brother, who had done nothing but carry out his work (ensuring the well-being of his patients). On the other hand, it is due to the secrecy surrounding the father's murder. Kader asks himself: "I wonder if the country isn't paying the price for the monstrous acts it committed in the past for the sake of a just cause."[16]

This novel brings to light what we can call a genealogy of fratricide since the War of Liberation. Colonialism, with its very real impacts, was used to mask the atrocities suffered within and by one's own community. When Kader learns of his death sentence pronounced by his brother, the narrator takes us into the private sphere of the condemned:

> He didn't understand this strange and perverse desire that pushed so many brothers and neighbors to kill their brothers and neighbors. He suddenly realized that a terrible monster was rising out of the abyss and that it was going to destroy everything. He felt that, despite his age, he was only a survivor. ... Is there a place for love in this world torn apart by discord?[17]

Fratricidal politics mirrors perfectly the bloody divisions between real brothers.

In Mansour Kedidir's novel *La Nuit la plus longue* (2015), we witness the despair of one of the many fathers who is told by the police that his son has died. As his private drama collides with the family war ravaging the country, he is overcome with anguish:

> "Your son has just died." I kept my cool. I had thought that Baroud had been taken down by law enforcement, dead without a doubt, since he himself, having sought this, enraged, became the specter of death. "I knew he'd end like this," I consoled myself. "No, I'm talking about your son the soldier, Rahim, he just died for his country." I didn't ask how he died. Because as soon as you receive the news, you're flooded with pain and you don't have time to ask this question. Only after the fact do you search for an impossible truth that in theory would bring a sense of calm. I remember that I didn't realize when I started walking around in sobs. After a little bit, I was running and screaming. Then I would go into the city yelling. I wasn't able to stop. People were running after me. When I reached the gate of the barracks, and I don't know why I went there, soldiers stormed at me and stopped me in my tracks. I was given a sedative, but I continued to tremble and drool.[18]

The despair and trouble of this father who asks himself which of his sons was killed recalls the major questions everyone in Algeria was asking during the years of the Internal War: "Who is who?" and "Who is killing whom?" Indeed, the fake roadblocks where Islamists donned military uniforms (after killing soldiers *en masse*) put the larger population on edge. The impossibility of distinguishing between the killers and those protecting from the killers sowed terror. The breakdown of the barriers separating terror and security was a central component of this war. The resulting confusion was used to further spread terror.

The question "Who is killing whom?" was often answered by blaming the state for countless targeted atrocities and large-scale massacres, some of which, at the end of 1997, took place near military barracks. Entire villages experienced indescribable horror for several hours: babies, children, women, men, and the elderly were slaughtered, their throats slit, their bodies cut to pieces and eviscerated (sometimes even having their viscera and heads hung from electricity poles so that everyone knew of the coming terror). Many works have spoken about this episode, especially accounts by victims and disaffected soldiers, shedding light on the despair experienced by individuals living in this state of profound confusion. It seemed as though the institutions designed to protect and defend the population had turned into killing machines.

This terrible Internal War was thus the scene of a profound confusion where political atrocities, the settling of scores, Islamic terror, corruption, and delinquency were all mixed together. Trouble and confusion often served as invisible shields, either for the political establishment or for regular individuals, who took advantage of the confusion to advance their own agendas unseen. In spite of this, there was an unmistakable specificity to Islamist terror, which was unleashed against the larger population. Nor was there any confusion concerning the large number of victims and the earnest struggle of many men and women who engaged in what they thought was a just fight. Indeed, the appearance of two warring factions – the state against the Islamists – is a gross reduction of what really happened: civil life in its entirety had been taken hostage by an internal war whose aims and limits remained unknown. Confusion hindered one's ability to conceptualize it, name it, and, worse still, to know where one stood within it, no matter how one felt about the matter. The confusion was so extreme that even today it is still hard "to believe" one theory over another that seeks to explain the war. People continue to change their position and adopt another line of reasoning to explain it. Understandings of the war have also been reduced to two competing theories. The first puts all the responsibility for the massacres and other destruction on the Islamists. The other insists that the state is responsible for orchestrating the war. This debate appears to remain unsettled even among historians.

Since the 1990s, countless reports and accounts published by Algerian and international human rights organizations have provided support for this second interpretation, without, however, tipping the scale in this direction. Indeed, this debate is caught in another type of division, lacking a thorough historical analysis based on a recognized set of facts and suffering from the absence of any judicial proceedings. The failure of the judicial system in Algeria to get involved as an independent body has only reinforced these problems and done little to dispel the terror surrounding the question "Who is killing whom?" Each interpretation offers some partial clarity and makes up for what is missing in the other. The one that sees the state as responsible has the advantage of bringing to light the hidden side of the political order, which, however, hardly ever hid its savage practice of fratricide. And the interpretation that accuses the Islamists of being the only responsible party for the war shows how political Islam transformed into militant Islam and calls attention to the failure of institutions and the political order, not to mention the resulting widespread destruction. Islamist terror has been appalling in a state that sees itself as primarily religious in light of the LRP. In this case, too, the debate is caught in a strict binary with no third option that could provide an unbiased account of who is responsible and which institutions have failed. Each subject is left to cope with the haunting questions of "Who is killing whom?" and "Who is who?"

These haunting questions produce fear and turmoil within subjectivities. They indicate the level of fear, terror, and confusion at work. Indeed, the endlessly repeated "Who is killing whom?" injects fear into social relations, provoking a feeling of vulnerability and reinforcing the belief that there is no justice. Citizens live their lives besieged by terror. They aren't so much concerned with definitively determining who is responsible between the state and the Islamists, because they know that both are colluding on an invisible level. On the one hand, Islamism was a product of the political order's aim to subjugate its people. On the other, Islamism made up for the failures of a political order too concerned with anything but the future of its people.

Citizens are concerned with the absence of justice and historical context, which only reinforces the terror they feel. The question

"Who is killing whom?" can also be understood as a call for erecting barriers between one another. Lacking justice, there is a frightening shift from focusing on the perpetrators of the crimes to a widespread suspicion not attached to anyone in particular, which confirms the belief that state terror and Islamist terror are working together on some level. One transforms into the other amid a merciless war. The debate won't be settled by eliminating one faction in favor of another. Hence the need for a dialectical approach that would bring to light the complexity of the stakes and facts of the matter. This can only happen when each institutional agency assumes its role in this enormous undertaking, which is still awaiting action. The general public has an essential role to play in this matter, but it still struggles to form a collective whole, as it, too, is plagued by suspicion while subjects continue to doubt one another. Just as under colonialism, the divide between "us" and "them" is at the heart of the Internal War. It shows once again that it is a matter of two similar entities fighting one another. An Internal War decked out in the garb of the War of Liberation. This divide masks the system's commanding power and the homogenization of various positions, which repeatedly prevents a third option from emerging.

The war comes home

The Internal War of the 1990s was therefore clearly a war fought against one's own people. The War of Liberation was resurfacing in another guise. The same signifiers and gestures were being adopted, in language, in the imaginary, and in countless actions, but this time the division was between the political order and the Islamists. With an Islamist Republic, the Islamists promised to abolish – through the simple and pure act of erasure – all traces of colonialism. For this, they deemed themselves "*mujahedeens*."

Those fighting for national liberation were also called "*mujahedeens*." The word *djoundi* (soldier) fell out of use and was steadily erased from memory. In other words, the War of Liberation slowly became "Islamized." The new "*mujahedeens*" (Islamists) found themselves once again fighting colonialism, but this time around they deemed the former "*mujahedeens*" (national combatants who fought against French occupation) the new colonizers. If one adheres to the use of these signifiers, it *almost* looks like a war between the new and old *mujahedeens*. In Algeria, the Islamists called the FLN and the "democrats" supporting those in power *Hizb França* (the party of France), aligning them at times with the colonizers, at others with the "*harkis*." Nationalist militias were formed in many regions across the country to fight the Islamists and to help people in villages who were taken hostage. For their part, these militias, frequently made up of veterans from the War of Liberation (*mujahedeens*), called the Islamists the sons of "*harkis*" ... The confusion of "Who is who?" reached a new height. Moreover, the Islamists were dying as *chahid* (the religious signifier of the martyr). But this word was also applied to fighters who died during the War of Liberation. A real war of martyrs! The "place de Martyrs" (*sahate el chouhadas*), in the center of Algiers, was incidentally where the Islamists regularly assembled from 1989 to 1990. The name of this plaza commemorated the memory of those who fought in the War of Liberation. From name to place, History and its subjects cannot shake the label of martyrdom. And let's not forget that the massacres of villagers between 1996 and 1998 took place at the major battlegrounds of the War of Liberation (Mitidja, Médéa, Mascara).

The rumor – or reality? – that soldiers, following state orders, were behind these massacres draws on two parallel arguments, which had already been put forth during the War of Liberation in regard to the massacres of villagers (FLN/French army). The state was believed to be behind and/or complicit in these massacres, since these villagers were thought to have voted for the FIS and were mostly in favor of the Islamist movement. In this view, their execution was an act of vengeance committed by the state against the Islamists. The other argument claims that it was the Islamists who attacked these villages, accusing them of not supporting their cause and of being "*harkis*" of the state. Now some notorious Islamists were believed to be actual descendants of "*harkis*,"[19] joining the Islamists' rebellion with their young brothers (this is the case with Hassan Hattab). The same geographical location was the site of horrendous atrocities during both the War of Liberation and the Internal War, forever troubling the way the site would be remembered. As the anthropologist Abderrahmane Moussaoui has suggested:

> Maybe there is a connection that needs to be made between the geography and history of these regions where violence has rained

> down. Indeed, these regions have witnessed violence since the beginning of colonization: Médéa, Tlemcen, Mitidja, Mascara, Sidi bel Abbès – all these cities located at the foot of mountains, along the Tell Atlas chain, were home to historic rebellions during the War of Liberation, which armed Islamist groups revived during the 1990s. Generally speaking, the rebel groups that fought during the War of Liberation were called upon again during this new war. To what extent does the location, this site of collective memory, channel and normalize such behavior?[20]

In its fight against Islamism, the state opened countless detention camps in the desert where people were routinely tortured. Deportation and internment camps had also been created by the French army to fight against nationalist fighters who were deemed "terrorists" by the French state. The fight against "terrorism" therefore borrowed from the French army's repressive tactics used during the War of Liberation. As for the signifier "terrorist," each period has its own, depending on who is in power: nationalist fighters were deemed terrorists by the French state and then later Islamists were given the same label by the Algerian state.

It is also worth pointing out that the FIS, after canceling elections and dissolving its party, split into several violent factions, made up of small, warring groups of brothers (AIS, GIA, etc.). Many Islamists were thus taken down by their brothers-in-arms. For its part, the state hoped this seemingly incurable "animal-like rage" would kill itself.[21] Let's recall that, during the War of Liberation, the French army's preferred war strategy was to encourage self-annihilation via fratricide. In post-Independence Algeria, the warring clans of "brothers" among the state were mirrored by an FIS that fractured into groups of brothers who began killing each other. As Moussaoui explains: "Photos that showed the crackdown by Algerian soldiers during the October 1988 youth uprisings alongside the crackdown by French soldiers during the pre-Independence insurrections were put up in neighborhoods throughout major cities during the electoral campaigns of 1990. The caption, which was hardly needed, read 'What's the difference?'"[22]

Islamist discourse and practices also emphasized the process of reparation of the father who was attacked under colonialism. The Islamists sought to reinstate a paternal function that could not only restore the father but also make him invincible. It was a matter of replacing the father with a powerful and vengeful God who existed outside all lineages. In place of the traditional father, the Islamists offered a greater, unassailable father: God.

Would the FIS [an acronym that notably rhymes with *fils*, or son] have taken it upon itself to save the brothers among the state from their own fratricidal war by reinstating the symbolic taboo as an order from God? Or would fratricide come to a definitive end through the election of the single-party FIS [*l'unique FIS*, or "only child"]? In both cases, the FIS ["the son"], in line with the aims of the LRP, sought to re-establish the father. Both the FIS and the LRP are concerned with the collapse of the paternal function under colonialism.

Depending on whether one uses the term "brother" or "son," lineage is seen from a different angle. The son has always been viewed as a descendant who gives life to his ancestors. He is thus the product of a line of fathers in a patriarchal society. In contrast, "brother" moves in a horizontal direction, which Freud says is the most susceptible to the lure of incest and murder.

In Algeria, the FIS ["son"] refers to the Islamic organization known as the Muslim Brotherhood established in Cairo in 1928, whose success was most likely due to the fact that many Muslim countries were occupied and colonized at that time. The Muslim Brotherhood's aim was to fight colonization and to permanently restore the major symbols that had been destroyed. Even today, Islamist militants, whether or not they are armed or engaged in *jihad*, call each other "*el akh*" or *brother*. The internecine wars that have plagued them are a logical reversal of this display of fraternity. The fratricidal fighting tearing the Islamists apart is not unique to Algeria. It bears repeating that these movements in several Arab countries were spurred on by the fight against colonialism. Let's recall that all the Muslim countries were colonized by the French or the British, with the exception of Turkey and Iran. This point can hardly be overstated given that these former colonies are all, in one way or another, currently plagued by what Fethi Benslama has called "civil wars": "The Muslim world," he writes, "is experiencing today a widespread civil war fought over what it means to be Muslim. What I mean by that is that individuals who recognize themselves by this term are fighting each other over the ethical and political understandings of the term's definition."[23]

A strange reversal in naming

The undermining – or even full-scale destruction – of the paternal function took on a new level in the inverted naming system used by the Islamists. Note that they would change their name as soon as they joined the rebellion. The same practice is followed almost everywhere today among self-proclaimed jihadists. It draws on a practice called *kunya*, which originated in the Middle East and consists of erasing the patronym and replacing it with a very specific name. In Algeria, colonialism also abolished the traditional naming practice, which had the father at its center (Ben, son of…), by setting up its own patronymic system of naming (via the civil registry).

The practice of *kunya* represents the return of what had been repressed from this history of the name and its erasure under colonialism. The Islamists named themselves by reversing the place of fathers and sons. They did this both by flouting tradition and by ignoring the patronym issued under colonialism. *Kunya* uses the prefix *Abû* (or *Abou*), which is to say "father of …," followed by the first name of the first-born son, or in rare cases the first-born daughter. And for those with no children, they draw on the first names of the Prophet's companions and/or the first names of the conquerors from the Islamic conquest. In Islamist reasoning, it is the son who creates the father, in contrast to the traditional system, where the father designated and lent his name to his son. In this way, reference is no longer being made to colonial or national history (with a state-issued patronym) or to the traditional system by way of previous generations (fathers, grandfathers). On the other hand, all of this points back to the site of the repressed signifier: the father, who in his symbolic function named, distinguished, and welcomed the being born to this world. With *kunya*, the father is given his place by the son. In this process, the son allows the father to exist (*kunya*), and not the other way around, as with the traditional system. The Islamists are still seeking the father who disappeared under colonialism and who, for the brothers in power, still can't be found.

The history and practice of Islamism (armed or not) uncovers a few truths about colonialism. As it strives to bring back the father in order to put an end to the fighting between the brothers in power, it shows the level of destruction wrought by colonialism on lineages and genealogies by making the father disappear and turning the son [*fils*] into the son of no one: the FIS. The Islamists sought to restore the father via the son by reversing the usual pattern of recognition which moves from the father to the son. *Kunya* allows one to move from the son (FIS) toward the father, and therefore to turn the missing father into a potential ghost. Better still, the Islamists put into power a FIS that is the son [*fils*] of God, at the risk of blasphemy since, in the Muslim faith, it is said that God is neither begotten nor has begotten any children. Seeking a place for the father, they erase him altogether in favor of God.

The Islamists were well aware of where the pain from colonialism was most felt. This explains why the general population, which, once again, found itself living as "*indigènes*" in the eyes of the political order, was ready to embrace them. The word "*salut*" [*salvation*], which appears in the French translation of the *Front Islamique du Salut*, could just as easily be translated by the word "rescue." One could thus call it the FIS's rescue mission of the father. Amin Maalouf put it best in his book *In the Name of Identity*: *Violence and the Need to Belong*: "You could read a dozen large tomes on the history of Islam from its very beginning and you still wouldn't understand what is going on in Algeria. But read 30 pages on colonialism and decolonisation and then you'll understand quite a lot."[24]

The goal of framing the Internal War in Algeria as another colonial war was clearly articulated by the Islamist Ali Benhadj, co-founder of the FIS. This war was haunted by the colonial war and was fought between brothers and only children in the form of elect sons (FIS). In 1991, Benhadj exclaimed:

> If my father and his *brothers* had physically expelled France the oppressor from Algeria, then, along with my brothers, my weapons of faith, I devote myself to the cause of banishing France intellectually and ideologically, ... so that Algeria may once again reign over the Mediterranean, in the form of Islam put forth by the Qur'an, the Sunna and the exemplary life of its pious ancestors.[25]

One of the leaders of the GIA (*Groupe Islamique Armé*, formed after the collapse of the FIS) declared: "The world must know that all the killing, massacres, fires, the uprooting of people, and kidnappings of women are offerings to God."[26] A very peculiar God who loves live flesh and blood. As a consequence, the more horrific the act, the more the Islamists believed that they were carrying out their mission to God. Hence their use of cold weapons (knives, saws, axes, etc.).

Terror was a veritable weapon of war, and all the more so given that, since Independence, Algerians had lived in a state of fear orchestrated by a military regime that, though working behind the scenes, wielded immense power. In other words, the reign of terror was propped up by the spread of horror via its visual display: dismembered bodies strewn about the streets for all the living to see; severed heads sent to the victim's family or suspended from electricity poles; the viscera of babies hung in garlands; newborns put in ovens; women and young girls raped and sodomized in the presence of their close relatives ... For its part, the political regime created fear through atrocities carried out in the shadows, including numerous assassinations of politicians and intellectuals, "forced" disappearances and torture, none of which has ever been acknowledged. The unspoken order was to keep these atrocities shrouded in silence and to fully exercise one's power in the dark. Thus two practices of terror and fear were at odds: that of the Islamists, which indulged in spectacle and the blatant visibility of terror, and that of the political regime, which favored a diffuse and muffled fear spread by invisible means.

The uprisings of October 1988 were fueled by a desire to move past fear through a democratic system and to bring an end to a form of governance driven by theft and fear of what was to come. And then right when a democratic form of governance was established for the first time in post-Independence Algeria, a state of terror struck again and spread, leaving people almost nostalgic for the fear they felt before. History is a strange phenomenon, as it is endlessly upturned and repressed, only to return again. What happened after the uprisings recalls what happened right after Independence when a revolutionary regime that had fought bravely against fear to achieve freedom once again instilled fear in individuals. What to make of the tragic fate of freedom witnessed in 1962 and again in 1988? And what can be said about the strange bedfellows of freedom and terror?

Do freedom and terror go hand in hand?

Since 1830, countless Algerians have been struck with some form of fear or terror right when a long-sought-after-and-fought-for freedom has been achieved. This reversal has profound consequences for psyches, bodies, and social relations. Algeria's political, religious, and linguistic history offers an exemplary case study for understanding democracy via its limitations and gaps. The emergence of democracies and their abrupt reversals encourages us to look back at the history of *demokratia* and its first manifestation. Let's not forget that the birth of the French colony sprang from the tension and reversals between a republican and a monarchical system.

Democracy invites and finds a place for what is different and dissimilar, making this an integral part of the social fabric in conformance with the ideal of social harmony.

This is what is meant by fraternity, which also entails freedom of speech and equal voting rights. This Greek invention welcomed (non-violent) conflict within the political sphere. There is no need to kill each other once everyone agrees to the democratic pact.

As we've seen, in Algeria, the "democrats" were caught in the trap set by their own demands for democracy, finding themselves divided and mired in two versions of totalitarianism. One concerned the political regime's totalizing agenda. The other was advancing a radical form of totalitarianism. Many Islamists took up arms right when, in the name of democracy, free and independent elections were canceled. Internal war arose out of the collapse of the democratic pact. This has forced us to rethink the relation between war and democracy. The democratic pact is designed to prevent an internal war by accommodating difference. But the canceling of elections in January 1992 never allowed for the sort of conflict this implies. This episode of political history illustrates how democracy and civil war are bound together almost like father and son.

In *The Divided City*, Nicole Loraux undertakes a meticulous study of the birth of democracy in Athens. She explores how the formation of democracy is born out of a need to escape civil war, and, more importantly, to move past it by forgetting it. Forgetting is thus at the heart of the political life of the city. However, forgetting is always wrapped up in what can't be forgotten, what is *impossible to forget*.

For Loraux, civil war (*stasis*) continually threatens democracy. Their historical concurrence will also be blanked out. Thus, the political order of the city is written via erasure. This writing of blank space nevertheless ensures that the specter of *stasis* looms everywhere and can at any moment surge forth in a display of blood and violence. In this way, democracy can be seen as a form of treatment for what Loraux calls the "divided city." According to her, voting is a "preventive remedy for bloody division" and civil war is "the unavoidable recurrence of evil on which the city is founded."[27] Under what conditions does conflict not lead to what Loraux calls "division"?

The boundaries between democracy and civil war are porous. At any time, one threatens to become the other. This reversal happened in Algeria when the general public sought to escape tyranny via a new system of governance. Fratricide thus constitutes both the collapse of politics and its foundation. The killing between brothers represents the bloody side of the formation of a political order, and this isn't unique to any one place. A closer look at the myth at the center of Freud's *Totem and Taboo* – with careful attention given to the footnotes – reveals this dark side of the Father's formation. Let's recall that, for Freud, fratricide can be part of this process. The shift from fraternity to fratricide is written into the history and formation of politics. The fact that fraternity is bound to the repression of fratricide suggests that the latter is primary. Whether *demokratia* or *stasis* emerges depends on the swing of the pendulum. Consequently, there is unfortunately nothing incongruous about these two dimensions of politics (*demokratia/stasis*). More precisely, there is a devastating continuity linking the two together, joined at times simply through repression, and at others by a reversal: as Loraux makes clear, democracy is promising insofar as it wards off threats. It strives to offer a path out of fratricide – one founded on equality and active participation, which calms the desire for exclusive possession.

Civil war, internal war, war waged against one's own people – all of these represent, in part, a mode of social incest as distinctions/barriers/separations all fall apart. Jolted by a profound sense of insecurity, the subject, caught between paranoia and melancholy, faces two options: to kill the other or to kill itself. The same matter is at play, which at times, in an act of reversal or subversion, leads to perversion. The political order erases the past, History is denied, and the living become haunted by a specter that takes shape before them.

Loraux believes that forgetting is at the origin of the unified city. It is a matter of forgetting the divisions that spurred on war and murder within society. The political order commemorates massacres by establishing citizenship, which is to say, a fraternity that is a reversal of and a deviation from fratricide. And at the same time, the political order also depends on a phenomenon at odds with forgetting: namely, erasure. The political order reshapes the very foundation of History so that the city can endure. The feeling of a "national" identity is therefore profoundly fratricidal. The power of murder is draped in denial by the political order and channeled into other forms of expression, hate of the foreigner being one of the most pronounced. This is perhaps why each society must designate the foreigner as marking its outer limits. It is a matter of protecting the city from the ravages of internal fighting. Hate feeds off of clear divisions (self/other, familiar/foreign). Today in France, as history repeats itself, Muslims find themselves again to their own dismay as the foreigners they were under colonialism. Note, too, that the specter of civil war even loomed over French society after the attacks on its soil by Islamist brothers.

The political order allows two modes of psychic life to co-exist in perpetuity: repression and foreclosure. These two mechanisms work together. They are brought together by denial, which is essential to the formation of collective history. Forgetting, which risks being reversed at all times, may trigger erasure, and vice versa. Reading memory must therefore always take into account its "immemorial" side.

Fratricide can be repressed and become fraternity, and/or it can be foreclosed, which gives it the power to act and to haunt society. The fratricidal origin of the political order can re-emerge at any time in a bloody and terrifying spectacle. Certain historical, political, and linguistic conditions influence whether fratricide is subject to repression or foreclosure, depending on what is known of the father. The argument that colonial violence undermines fundamental taboos, and, consequently, the symbolic world of subjects (namely, the paternal function), accounts for the impact, on society and subjectivities, of a law governed by chance rather than one designed to guard and protect. In Algeria, the Internal War of the 1990s was the product and blood-stained commemoration of the War of Liberation. As colonial divisions gave rise to isolated communities, it wasn't long until those same communities turned on each other. It is this transition that led Nabile Farès to write: "I will have to think. That's it. To think and to find out what history is mine."[28]

It is worth considering in greater detail the state of terror occasioned by the Internal War, especially regarding its impact on subjectivities and on thought, which came under increasing strain at the time. There is a long history of fear and terror in Algeria. They go hand in hand, but, as we will see, if fear allows the subject to maintain a certain connection with its emotions and thoughts, terror, for its part, seizes subjectivity in its entirety. Terror thrusts subjects into a world without sound, dreams, or shelter.

7

State of Terror and State Terror

The captain hadn't forgotten what he had learned in military training: his duty as a soldier was to protect and defend the country from all enemies, whether they be interior or exterior. Now that they were shielded from the foreigner by seas and oceans, and the era of conquest was over, he was convinced that the country was invincible to any exterior enemy. ... But on the other hand, the interior enemy was right there, impossible to grasp. He couldn't identify him. He has the same physical features, wears the same clothes, breathes the same air, eats the same food, and lives in a similar dwelling: to point his gun at his brother was like turning it on himself.

Mansour Kedidir, 2015[1]

I couldn't understand how people born among us, who were said to share "our home," could commit crimes in front of us, and not just any crimes, but cutting off heads, arms, feet, hands, noses, ears, fingers, as though they had inherited all the world's crimes and this were their filial duty.

Nabile Farès, 2010[2]

The official death toll of the Internal War is 200,000, not including the countless scholars, executives, and students who fled in exile. Algerian society suffered a massive loss of people. The journalist Hassane Zerrouky has counted 11,000 acts of sabotage and destruction carried out by the Islamists on various state institutions. He puts the cost of the war at around $20 billion, "or the equivalent of two years of oil production and several hundred thousand salaried positions furloughed."[3] Algerian society found itself as a whole in a severely weakened economic, political, social, and demographic position.

Although it may be possible to quantify the material damage of the war, it is very difficult to identify and catalogue its subjective effects: the experience of catastrophe, the hollowing out of existence, and the breaking of social ties with the Other, not to mention the consequences of all this on the social fabric, the pervasive feeling of insecurity, and its corollary, denial. Social cohesion was severely undermined as the very notion of social harmony became unthinkable. The work of Francophone writers hardly sparked private or collective revolutions. It remained, in spite of its relevance and social commitment, relegated to the realm of literature. When the far-reaching implications of these works were perceived, they were distorted for personal gain.

In Algeria, after these many years of war, religious zeal started to wane and a form of secular thinking began to emerge. However, in an odd turn of events, what could have led to a separation of religion and politics in civil society turned into its very opposite: a resurgence of religious zeal within discourse, with religion dramatically reduced to a narrow moral code. Spaces for thinking, connecting with others, and living were pervaded by this understanding of religion. This is all the more troubling in that this understanding of religion was drawn on to treat the feeling of terror provoked by the countless catastrophes experienced over the years. Out of the fear of God (an expression literally translated from Algerian Arabic) emerged a God who spreads terror, only further fueling the nightmare that had rattled the social body. How can terror – with its many points of convergence and divergence with trauma – be understood in clinical terms? Does terror require collective treatment?

A clinical understanding of terror

Terror is related to fear, but works on another level. It goes deeper than fear, anxiety, and maybe even dread, while intermixing with these three affective states. It is not solely an emotion, but also a psychic state. Terror doesn't reveal itself through speech or other telltale signs. It lives in the body of unwitting subjects. Without warning it seizes physical bodies and, at the same time, in the same indistinguishable impulse, the social body as well. The threshold separating the individual from the community, the body from the psyche, words from actions, is irreversibly dissolved.

The subject and the social merge to such an extent that they become almost indistinguishable. The dominance of censorship in the speech and thought of patients in Algiers attests to a psychic life that obeys a state of terror. At the same time, it seeks to create a safe zone, in a separate, private space. If it is easy to perceive this obedience to censorship at work, it is nonetheless difficult to make out its resistance, which remains carefully hidden, but undeniably present. Its resistance constantly escapes perception, eluding any attempts to seize it and fix it. Psychic denial, a central mechanism in the formation of collective history – no matter whose history – produces various forms of censorship, which work to keep the writing of the political text illegible. The Internal War in Algeria was unleashed against those whose weapon is the pen.

From 1993 to 1998, many Francophone (though not exclusively) writers, journalists, and intellectuals were targeted, shot down, or cut up. Armed Islamists waged an all-out war against the pen. In a 1994 communiqué, the Emir of the GIA, Djamel Zitouni, declared: "He who fights us with the pen will perish by the blade."[4] The statement is clear and unambiguous: ink spilled on paper will be replaced by the spilling of blood. The murders that ensued were numerous and especially savage. The list of writers killed is long (more than 100 between 1993 and 1997): Tahar Djaout, writer, journalist, and essayist; Abdelkader Alloula, playwright and director; the psychiatrist Mahfoud Boucebci; the sociologist M'hamed Boukhobza, his throat slit before being disemboweled in front of his family; Ahmed Asselah, director of the School of Fine Arts of Algiers, murdered at his school at the same time as his son Rabah; Vincent Grau, head of the Fine Arts bookshop in Algiers; the poet Youssef Sebti, as well as many artists ... A major portion of the intelligentsia annihilated.

These murders remain today plagued by the same question: "Who is killing whom?" Indeed, the majority of these intellectuals had fought for free thought and against the political order's mode of governance. They were also firmly opposed to Islamism. Even today, with the absence of trials – a subject to which I'll return – that would name the assassins and reveal the complex history of these atrocities, the question "Who is killing whom?" continues to work in favor of the censors and to fuel a sense of insecurity. The traces from this process are not only illegible but also are not even documented. More precisely, these traces are unwelcome, as the mere acknowledgment of their existence would plunge many subjects into a state of disavowal of their own insecurity regarding the truth of the facts. In other terms, the erasure of traces has become a state function performed in each individual without distinction.

Before his execution, the writer Tahar Djaout had written in one of his columns what would later become a slogan for the independent press: "If you speak, you die, if you don't, you die, so go ahead: write and die!" He was indeed killed in June 1993. Who killed him, as is the case with so many other murders, will remain a mystery. These intellectuals were massacred while the perpetrators of the crimes, replacing ink with blood, remain anonymous. With no hope of anyone being brought to trial, can there ever be an end to the Internal War? How can the living find closure after these savage crimes if justice remains to be written? This Internal War is a war over writing. Everything that is written is erased by blood. All must obey this diktat: make all legible traces disappear in favor of a display of mangled bodies. In this way, death itself, more than the dead, becomes disfigured. This brutal logic reigned for many years and redefined the relationship between the living and the dead. The cruelty scrupulously exercised on bodies, regardless of age or sex, was addressed to the living so that they would know their own fate. It was also a message sent to death itself. For dispossessing the dead from the integrity of his or her body by cutting it to pieces is as much a violation of life as it is of death.

The death of the individual isn't enough for the killers' ruthless logic, which strives to go beyond death through disfiguring, hacking, and opening up bodies, not to mention hollowing them out from within. This is what the psychoanalyst Paul-Laurent Assoun has called "killing the dead."[5] He was referring to the French revolutionaries who, in the aftermath of the French Revolution, and in spite of having already killed the king, felt the need to reopen the royal tombs in order to remove the bodies and hack them to pieces. In this way, the body was dispersed into a multitude of parts, defying reassembly. In France, the Reign of Terror followed the Revolution, in this strange stage of moving beyond the dead to attack and disfigure death itself.

The treatment of bodies by the armed Islamists aims to de-sacralize death. With this they reach a new level of blasphemy since, to use their own terms, it appears that the body of the dead is no longer the business of a single and unifying God, but their "own" business. The deliberate destruction of the corpse is a way of spreading terror through the social order, which, for the living, is seen as irreversible. With this level of slaughter, bodies are so unrecognizable that the living can no longer even speak of murder. They are thrust into a world beyond death. In light of this, individuals informed of the death of someone are haunted by a terrifying question: "How did he or she die?" – the understanding being, is the body in one piece? If it was a question of a "natural" death (sickness or the like), then there is a tangible feeling of relief. "Natural" deaths brought a sense of peace. This is tied to the fact that the dead hadn't yet disappeared from the world of the living and that others could also be met with this reality. Which raises the following question: to what extent does terror exist outside of the real?

Discussions from this period never showed any trouble distinguishing between death and that which worked on another level entirely, where, for instance, the body of the dead found itself excluded from death. The living felt this disfigurement of death in their own bodies. Terror disables the regular circuits connecting the body and the unconscious. This explains how thinking can become policed when the pathways connecting thought to an organic body have been cut off. With terror, an otherwise inaccessible zone of the body and the psyche is reached. The terror reigning outside is unwittingly absorbed to become a persecuting inside. As the boundaries dividing the inside from the outside disappear, a logic of persecution takes hold whereby the persecuted and the persecutor imperceptibly merge together, raising the famous question: "Who is who?"

With the breakdown of regular filters, the atmosphere and the mood from without have set up permanent residence within. This works by obliterating the internal divisions separating inner and exterior reality, body and psyche, persecutor and persecuted. Terror mixes together spaces that are usually distinct and held separate. The terrified subject can no longer identify why it is being besieged or by whom. Its thought circuits are blocked. Thus vanishes the ability to designate a persecutor, which is to say, the one who is responsible. The subject is confronted again and again with the question "Who is who?" Now if one cannot recognize the perpetrator of this state of persecution, then one's body turns on one's self and becomes one's own persecutor. The outside atmosphere mirrors the ruthless attack from within, making the question "Who is killing whom?" unanswerable. No name or face can be attached to an outside persecutor, nor can anything allow this state to be recognized as produced from within, in contrast to fear, as we shall see.

In this way, each person becomes, through his or her body, an authorizing agent of his or her own death. The individual reaches a place where body and psyche are one, reduced to a single, indissociable point. Body and psyche become the atmosphere and the space of dismemberment. Terror thus has a haunting quality, which, unlike those conditions that restructure the private sphere (neurosis, psychosis, perversion), can be described as a state of hallucination *in negative*. The besieged subject hears the threat of its own imminent death from an unattributable blank voice, belonging neither to it nor to the Other. It thus develops "hyperacusis," a pathology that makes it frighteningly alert to the slightest noise, which it scrutinizes and over-analyzes. This extreme sensitivity to the most minute noises and sounds shows how the subject is unwittingly and constantly on guard.[6] It must at all costs set up barriers and means of distinctions within its own perceptions, especially since it has no idea where these noises are coming from. It remains on the lookout, striving to spy a site where it may find protection from the feeling of imminent death.

The terrified subject's self-elimination

In his novel *Maintenant, ils peuvent venir* [*Now, They Can Come*] (2002), Arezki Mellal calls attention to this stifling atmosphere and the way it penetrates the subject's body. Mellal seeks to designate a space beyond death. He alerts the reader's attention to the inadequacy of the word "death" to describe an indescribable yet very real mood, but he is forced to use this term for lack of ... words. At stake here is a very particular death, the smell of a corpse-less death since the body has been hacked to pieces.

The name and the body become dissociated. Indeed, the corpse guards the name and re-engages with the process of birth where a name is attached to a living body. In all cultures, from birth until death, there is no body without a name and no name without a body. But in this novel hacked bodies have sent the living on a quest to reassemble the scattered pieces of their dead in order to bury them with a name and give them a semblance of a corpse. Many families remain nonetheless unable to give their dead a burial worthy of the name. One's body parts are mixed with those of others when not lost altogether, making the living "bone seekers."[7] Reflecting on this atmosphere, Arezki Mellal writes:

> It's as though death were everywhere. "Everywhere," now there's a word Algerians like to utter ... we'll talk for days about death, which has become an everyday reality, about the relentless killing, about the butchering of a human being in the cruelest way, all in the name of God. A murderer's nickname is "carpenter" because he cuts up his victims with an old rusty wood saw. For, they say, the more the victim suffers, the better the chances he will go to heaven. And the mission of God's madmen is to send us to heaven. Now tell me, Salah, how did it come to this? How could these people become like this? How could anyone kill with such cruelty?[8]

This savagery isn't limited to those who physically suffered from it. It spreads through psyches and the social order like a warning of what is to come for everyone. Mellal alerts the reader to this strange state of terror by letting the effects of a death sentence slowly take over the text. Which suggests that, with terror, it doesn't matter whether the sentence is real or hallucinated. In both cases, the same inescapable fate awaits the subject.

Fear gives way to certainty. One way or another, the person who is condemned to be executed and to meet a cruel fate knows just that this will happen, either by the hands of criminals, or, failing that, perhaps by his or her own, as we will see. *Maintenant, ils peuvent venir* puts on display the madness of the besieged. The subject knows they will come and, at the same time, its subjecthood depends on affirming a temporality. It is *now that they can do it* [*maintenant qu'ils le peuvent*]. In its own way, the subject sets the tempo, and the novel plays out on three levels.

Knowing the butchery will come is a matter of certainty. In Mellal's novel, written in the first person, the main character visits a communist friend because he knows the latter will be killed. The friend knows he is awaiting execution, even though he hasn't been specifically singled out, as had some intellectuals in Algeria. Both have no doubts about what will happen. At one point, the main character tries to convince his friend to leave the city and flee to a far-off location, telling him, "They are going to come, and you know it, now, tonight, tomorrow." The condemned man, in shock, awaits his "death sentence" (Maurice Blanchot); and the main character is left to say: "That's that, Salah won't leave, he is going to die, there's nothing that can be done."[9] The sentence has been pronounced and has irremediably sealed the fate of the condemned.

It is just a matter of time before it happens, and that time is just around the corner. There will be no surprise or enigma when Salah's wife, Baya, "will stumble upon a trash bag a few days later outside her door. She won't need to open it *to know*, the puddle of blood will say it all. She'll never recover Salah's body. All she'll have is his head in this bag."[10]

The first-person narrator is convinced of his friend's terrible fate. There remains no trace of suspicion or doubt. He will also in turn have to await his own coming execution without knowing when it will befall him. However, he isn't too worried about this, or at least on the surface. (He is not an obvious target of the Islamists as the communists were.) Still, he is sure of his fate. So he tries to escape his destiny by moving to another city and adopting new habits in order to elude the inevitable. And yet ...

And yet, although fearful of the roadblocks that could lead to his death at any moment, he takes to the road with his young daughter, Safia. His body is seized at all times by two feelings, between which he continually wavers: "anxiety neurosis" (Freud) and the certainty of his "death sentence," neither of which he can shake off. His daughter understands something strange is going on with her father. They're traveling on the route of the Mitidja (a plain right outside of Algiers, labeled the "triangle of death" during the Internal War). His young daughter tries to distract him while he is suffering from almost hallucinatory terror. The child keeps talking to him to keep him awake and to calm his state of terror so that he doesn't succumb too quickly to his anticipated death, which seems ready to strike. Safia thus tries to fill her father's imagination with stories to drown out the "blank" voices he hears in his head. But then, at nightfall, the car breaks down. There's no doubt, it's now that they can ... come.

The ignition isn't working. He is suddenly speaking about his car as though he were talking about his own body: "But I keep returning to the ignition and try to start the car. Irrational act of despair. The carcass makes only a heart-rending groan."[11] The slightest noise is heard and interpreted as what he and his child will experience in the next minute.

From there, the reader is led through a haunting trip where it is impossible to determine what is real: is he seeing (hallucinating) shadows, or are those murderous, savage shadows men who have come to kill him? Throughout, the text is pulled in two directions: toward an overwhelming sense of the real and toward an unreal that needs to be scrutinized, seen again and again: "I open my eyes wide ... I scrutinize my surroundings."[12] In addition to what he sees (real and/or hallucinated), there is what he hears. He lends attention to the slightest noise. But how can he know what is coming from within, tied to "anxiety neurosis," and what is coming from outside? And above all, how can he know if the absence of noise outside isn't simply the absence of noise from within, which is to say, the "death sentence"? And if the already dead body remains the sole witness of a death that had already struck *and* is still in the process of coming? He is no longer afraid because now he knows. He'll say: "I'm amazed at my composure. I'm not afraid anymore. Safia is in my arms. Howls tear through the night. A group is coming. Suddenly dizzy. An eternal moment has paralyzed me." It's Safia, trembling, who asks him: "Are the slaughterers struggling?"[13]

The father responds to his child, literally beside himself, in an extreme state of dissociation, his actions causing him to be in this state, he who will murder his daughter so that she may escape the slaughter, which either she will witness by seeing what happens to her father's body, or he will witness by seeing his daughter being slaughtered.

By killing his daughter (by choking her), he tries *almost* to recover something from death. Better to kill his daughter and escape the massacre that thrusts one beyond death than to experience the coming slaughter. This father who murders his child is also his child's savior in a world where murderous shadows, death and cruelty, love and death all mix together. The father cries: "A piercing thought runs through my mind, resolute: they'll never take her alive! My hand seizes her fragile neck. 'Papa, you're hurting me.' I squeeze with all my strength. Cartilage is quickly breaking under my grip. Her head falls on my chest. *Now, they can come* [*Maintenant, ils peuvent venir*]."[14]

He, the child's father, a murderer who shields his daughter from the brutality unleashed on bodies, transforms into "them," the terrorists. The father kills his child to avoid robbing her of death, a matter that belongs to the living. The hands that rock and caress the child's body are the same hands that strangle it to preserve its integrity. At least in this case, the body remains in one piece and the murderer has a name and a face. But the worst is yet to come. Before the book closes, Mellal describes, in a different font, the encounter with the Emir and his men who apparently witnessed the murder: "Let this dog live. Let him speak of what he saw. Let him speak of what he did."[15]

Here, the terrorists are the ones who witness the murder scene, and they can hardly contain their joy. They let the father go so that he may succumb to his own psychic torture. With the change in font and the strangeness of the whole scene, the reader is left wondering if this encounter with the Emir really took place or if a terror-stricken father hears and sees a scene that is both imagined and yet strikingly real.

When a death sentence – real or not – is pronounced, one has no choice but to act upon it. Under the reign of terror, between thought and body there is no translation, interpretation, or detour. Only imminent arrival. This helps explain the vast number of irreversible illnesses that turned fatal for many people during and after the years of the Internal War as the living became paralyzed by panic. Countless cases of cancer, strokes, and heart attacks raise questions about the full impact of the state of terror. Many men and women, knowing themselves to be under threat, died of brutal organic illnesses. Claiming that death surged forth in the bodies of the living is a difficult argument to prove, and yet it seems unavoidable. The state of terror lays siege to the body with a hallucinatory certainty: whether the death sentence is real or not, it acts as though it is. This raises a crucial question: what is it about terror that the subject can neither forget it nor return from it? How can this transition-less site of the psyche where repression seems absent or inoperative be understood? Is this what leads psychologists to consider the reality of the body – its presence in the real – as a site of the psyche prior to repression?[16] And could the body's realness become in certain conditions no more than the space where a death sentence is readily carried out?

Under the reign of terror, the death sentence is an order with which the body unhesitatingly complies. "Anxiety neurosis" induces physical and psychic agitation as a sort of bulwark. However, this resistance is no match for the certainty of the "death sentence." What can be described as inertia, the feeling of being stuck or not fully alive, seems to point to a reign of terror that is still in effect in Algeria, although to a lesser degree than during the war. Psychic censorship ensures that the state of terror is upheld by barring off certain zones, severely controlling circulation like a sort of internal curfew, which speaks to the crisis afflicting the living.

If terror reaches an unknown space of the unconscious psyche through its direct and unmediated contact with the body, then it is worth considering how trauma operates in relation to this. The besieged, terror-stricken subject's hypersensitivity to noises, sounds, voices, footsteps, gestures goes beyond a state of stupor which is related to sight.

Stupor is at play but it is more aural than visual. All the living's senses are in a state of panic, eagerly awaiting the realization of the death sentence (real or hallucinated). The aural dimension of terror is unique and much more pronounced than the stupor found in the hyper-alert gaze. An archaic space of the psyche is activated under terror which, in theory, remains inaccessible to the subject. Let's recall that the fetus's first contact with the outside world is through sound. Does hearing's inability to repress (forget) become in a state of terror the *impossibility to forget*, which remains active even during sleep?

Psychological terror is always political

After a requisite period of shock, trauma triggers the need to speak, to take stock of what happened and to provide an account of it. Trauma is a marked subjective experience. The traumatized subject tries to connect the traumatic event to the rupture it provoked. What's more, the subject can designate trauma – at times unknowingly – as an event, which in the best-case scenario allows for a new subject position to emerge as one faces the world anew. The Internal War, by provoking a state of terror within both the private and social spheres, caused many traumatic experiences for individuals. Nevertheless the notion of trauma cannot account for the way a "siege" takes place from within and determines the fate of the besieged.

From a clinical standpoint, I would argue that a state of terror lies outside the realm of trauma, as it doesn't allow for a new subject position for the living. Although its relation to life may be deeply affected, the traumatized subject doesn't die from its trauma, whereas the terror-stricken subject often does die from its terror. Clinical work has shown that there is a perverse pleasure provoked by trauma's ability to immobilize the subject. The subject clings to its trauma as though it were part of its own body, replaying the traumatic moment in an endless loop, to the point of making it a constant state of its living experience. The terror-stricken subject has nothing to cling to, since its foretold execution will soon happen by its own hands or those of its killers. The subject's body turns on it and becomes its murderer.

With terror, disappearance is one's expected fate. To take initiative, to feel alive, to take matters upon one's self – all of these appear as dramatic transgressions within the private and social spheres where terror is concerned. On the other hand, striving to lead an inconspicuous, "pared-down life" (Nabile Farès), by obeying the censors and the various diktats of the political order, seems less costly, whereas it is paid for by stolen dreams and a stifled interiority. In this case, the censors serve as the diligent guardians of a psyche besieged by, and destined to, death. Censorship, like palliative care, keeps an already internalized death active, by maintaining doubt and hesitation about the state of death.

In 2010, Nabile Farès provided an apt description of how terror lives on within the psyche when it hasn't already led the terror-stricken subject to its own execution:

> What's the difference between dying and remaining suspended between life and death in a no man's land of desire, joy, love, as though the incredible and freak fact of having just escaped death not through his own doing must remain engrained, in the form of this same death, in his head, as though a doubt of really existing pervades him whereas he had been completely spared in the attack, the explosion that left eight people dead – then nine, according to the papers, and many more wounded?[17]

Depending on the degree of the internal struggle, the radical certainty of death as expressed in the declaration "*Maintenant, ils peuvent venir*" (as in Mellal's novel) can turn into an "existential doubt," unsure of whether one is dead or alive. This underlying doubt leads the subject to second-guess the certainty of its death sentence. But it also reveals how the end of war hardly brings an end to the terror that continues to plague psyches and the social order. Death foretold (hallucinated or real) continues to loom in one way or another, until it either seizes hold of the body, or lets it live in this state of suspension between life and death, neither living nor dead. This suspension is a defense against having to experience "atmospheric death" (Fanon), or a sort of "living death." This is what Freud refers to with his principle of inertia. The subject is caught in a state of inertia that works against its drive to live. Trauma allows one to identify, isolate, and document the traumatic event. Terror works on another level altogether, since the savagery happened to someone else (and the dead don't speak). The perpetrator can't be identified (see below) and terror, for its part, cannot be localized but has clearly penetrated the very cells of the organic body.

What lies inside and outside of the self are perfect matches. In this context, the function of censorship is to re-establish barriers that were shattered. In so doing, however, it erodes the living potential of thought and speech in order to keep psychic life to a minimum. Once these barriers give way, the social order erupts in violence. Censorship keeps an eye on the living, perhaps to ensure it doesn't disappear entirely. Lady Psyche, consumed by terror, "hastens to bring together in a semblance of unity the various fragments that must be brought under control again."[18] Despite its viciousness, censorship paradoxically plays the role here of the psyche's protector.

The Hungarian psychoanalyst Sándor Ferenczi (1873–1933) offers many crucial insights in his writing that help us to understand the various levels of rupture experienced by the psyche. He considers trauma a plurality of expressions and translations. At the outer limit of trauma, which is to say, approaching what lies right beyond it, one reaches unknown sites of the body and psyche: "In moments of great distress," Ferenczi writes, "which the psychic system isn't capable of handling, or when these special organs (nervous and psychic) are destroyed in acts of violence, very primitive psychic forces are awakened and respond to the disturbance. *When the psychic system fails, the organism begins to think on its own.*"[19] Failing to conceptualize its inevitable death, the organism invites death in by making itself a welcoming site.

For Ferenczi, while the attack on the psyche is substantial, to the point of abolishing the subject's capacity for thought, the subject attacks itself from within so as to protect itself from greater destruction stemming from "immense distress, the danger of death, agony."[20] Part of the psyche is kept in a state of deprivation so as to remain localized and to avoid shattering entirely. Nonetheless, if the state of destruction is vast and this process is incapable of handling it, then the interior attack is all the more damaging and it is up to the organism, which is to say, the body as a presence within the real, to take over for the psyche. In this way, an unrecognized aspect of psychic functioning is exposed by which the living organism becomes the equivalent of the psychic unconscious. One is seemingly made perfectly continuous with the other, without undergoing translation or transformation. This is a very compelling argument, and it isn't surprising that Ferenczi made this discovery with agonizing patients whose psyches were overwhelmed by the feeling of imminent death, a feeling that emanated from beyond the realm of their anguish. Ferenczi never uses the word "terror," opting instead for trauma and "primitive death throes." But the description of the psychic state of his patients clearly indicates a state of terror. Except that a state of terror, as described above, is defined by its irreversible connection with the outside. A state of terror arises when terror is identical both inside and out. In other words, *this form of terror is always a form of state terror.*

Ferenczi discovered that some patients, exhausted from the relentless struggle and lacking any form of mediation between the inside and outside, can succumb to terror. The organism gives up, either through suicide, or by an equivalent to suicide, a sort of internal death, of which "autoimmune diseases" are but one consequence.[21] Autoimmune diseases spring from an abrupt reversal enacted by the body, whereby it is no longer able to see to its own self-preservation. The body gives itself over to death completely in search of respite from its internal agony. The weary subject "gives itself up" to find "deliverance."[22] Reduced to no more than the body's sheer presence – its existence as part of the real – this form of self-abandonment leads the waning subject to become its own murderer. In this way, just as in Mellal's novel, it spares itself from being slaughtered by the Other.

Terror is hard to keep in check. It resides at the juncture between the organic and political orders, and, without any visible traces, its effects are consequently infinite. What's more, its very functioning depends on an erasure of thought. Nevertheless, there is an urgency and real need to understand terror in clinical terms with the hope of finding some ways out of this dark silence. To paint a picture of this state, Ferenczi tells the following story:

> By way of analogy, let me tell you the story of a trustworthy Hindu friend who is a hunter. He sees a falcon attack a small bird. The small bird begins to shake as it sees the falcon approach and, in a matter of

seconds, flies directly into the open beak of the falcon before being swallowed up whole. *Awaiting an inevitable death* seems so difficult that, by comparison, real death comes as a relief.[23]

Terror leads to disappearance. Which makes it all the harder to grasp. One is left to wander through a blank world in a dark silence peopled by all-consuming specters. Terror seems to possess an almost absolute power fueled by a sort of pleasure it finds in death, a power that remains out of the subject's reach but resides in the presence of its body. Terror gives rise to a haunting and very specific world (since it is materialized). It consumes the subject. And it irreversibly tears apart the social order. When terror is exposed, it means that *the unconscious psyche is active in both the political and organic orders*.

Existing beyond trauma, terror cannot be treated solely as a matter affecting individuals. It also affects the social body. Thus, a form of care is needed that addresses both levels. Without providing urgent care to both the private and the social spheres, this death sentence threatens to take away many lives via death and/or inertia. A form of internal death is being tirelessly re-enacted without producing anything new.

Reconciliation: state terror?

The Internal War has been treated in Algeria by the law of *omertà*. Which, on the one hand, legitimizes murder and savagery, and, on the other, dismisses the entire affair. This has come to characterize political history in Algeria, and the impact on psyches has been significant. No work that has sought to explain this war – from a historical, sociological, political, anthropological, or psychoanalytic perspective – has had any major social impact. This is because the general public has yet to be invited to take part in this debate. Which has hardly deterred scholars and other professionals from carrying on in the exact same way, each in their separate fields. Their work remains unread. The majority of the population is left out of these debates, in spite of the fact that the state of terror spares no one. Terror is diffuse and pervasive, striking people both directly and indirectly. It disables one's distinction-making capabilities in order to keep the subject immobilized under the spell of dread.

In 1995, a law known as the "law of Rahma" (*kanoun rahma*), literally, "law of mercy," was adopted in order to find a solution to the war. A religious signifier used in political discourse in the context of a war fueled by religious zeal! This was followed by the adoption of a law known as the "law of civil agreement" under President Bouteflika. Inspired by the model used in Rwanda after the genocide, it permitted those who fought in the rebellion to lay down their weapons and rejoin civil society. Its aim was to restore the previously demolished hope for social harmony.

In 2005, this law became "the Charter for Peace and National Reconciliation" and a referendum was held over its adoption. The validity of elections in Algeria was again thrown into doubt by odd numbers. As always, this law was approved by an overwhelming majority (90%) while no efforts were made to engage the wider public in a substantive debate on the matter. This law of "reconciliation" permitted the Islamists to leave the frontlines without having to face any form of justice (articles 3–4, "Exemption from Sentencing"). Countless Islamists were released from prison with no further trials or sentences (articles 27–9, "Mitigation of Sentencing"). In theory, those identified as having taken part in the widespread killing of civilians and police and soldiers, as well as those guilty of rape, were excluded from this. The justice system was no longer involved in judging a subject's responsibility.

The initial plan was to have a commission made up of members of the general public and state representatives that would examine each case. But this never really materialized and the "rehabilitation" of Islamists was granted based on their own unverified testimonies. In the end, it came down to a testimony of faith. Let's recall that one becomes Muslim by proclaiming: "I testify that there is no God but God and Muhammad is his Prophet." A clear perversion of meaning is performed here, one that eludes any questions bearing on one's responsibility. Dismissals were granted on a large scale, with the extra bonus of financial compensation for harm caused to those "testifying," whose responsibility remained unquestioned. And so, in the same town, in the same waiting room at the doctor's office, victims sat side by side with murderers. Worse still, no effort was made to identify the perpetrators of the massacres and killings, which would have allowed the general population to partake in the formation of this history and cleared the way for a future mourning.

This law exacerbated the questions "Who is killing whom?" and "Who is who?" Nothing was done to appease one's internal sense of insecurity. And the reign of suspicion was further stoked among the general public. As a result, the censors had their work cut out for them, since this difficulty of discernment was a result of their decisions regarding what counted as acceptable speech and thought and what was forced to remain hidden within the body, under one's skin, so to speak, due to the widespread suspicion. Both the general public and scholars lack sufficient information to make sense of the Internal War, to be able to treat it. Many questions still remain unanswered, starting with: who committed these acts of savagery? And how did they manage to unleash this level of savagery upon the social order?

To the best of my knowledge, trials are extremely rare, and when people are tried, the general public is kept in the dark. How, then, can one even begin to develop a story, a history, a thought in this situation, and hope for terror to transform into trauma? Many taboos persist. More than 3,000 women were abducted and raped by the Islamists; some were even killed when they became pregnant while others served as slaves and gave birth to children in servitude: what did these women experience, what can they say about what they went through? No space has been created for them, or their children, or the general public, to tell their stories. Terror, taboo, and the fear of knowing continue to work in silence. How can one not worry that these state-sponsored dismissals are a sign that the worst is still to come? How can one not imagine that this rehabilitation only invites a re-enactment of history? Once again, the distinctions between assassins and victims break down and everything becomes hazy. What to make of the general public's exclusion from the process of coming to terms with, and, by extension, collectively reshaping, History in a country where 1.5 million "martyrs" gave their lives during the War of Liberation? (This is the official figure put forth by the Algerian government.[24])

And so, there are no living, there are no dead, neither criminals nor victims, no one has been found responsible or guilty. A law orders the pursuit of an orchestrated denial, which was voted for by an overwhelming majority! How to avoid thinking that this law is a form of state terror? And to what end?

The absence of justice, which seeks to abolish responsibility, gives a paradoxical message. Under the guise of rehabilitation, crime has been sanctioned, which afflicts the whole social order. As a result, the state of terror has an excuse to persist. Victims' associations have been speaking out against the injustice. They have found themselves once again victims of a vicious law that, under the guise of reconciliation, imposes silence and legitimizes crime. This law is a continuation of state terror, underwriting and supporting a widespread state of terror by authorizing the erasure of the war's catastrophes.

It is but a small step, and a couple letters, from amnesty to amnesia. With each war, the writing of History is dictated by denial, preventing it from being told in its "own" manner. This is far from the spontaneous work of forgetting, which serves as the foundation of collective memory. Here, foreclosure [*forclusion*] is the law. Jacques Lacan drew on this juridical term to signify the psychic effects of a signifier (The Name-of-the-Father) that doesn't get transmitted, and, as such, remains precluded.

What is retained via memory in Algeria comes bound with a preclusion that prohibits mourning and remembrance. Once again, the law dismisses memory. Make no mistake, this is the process by which colonialism was founded. But now it is practiced in the name of a free and independent Republic. The preclusion of memory gives birth to censorship, which aims to create the semblance of law, or, to be more precise, a caricature of law, as the law fails to uphold its purpose to protect individuals in the social order. However, this caricature of law only further pushes the subject to rely on a logic of *détournement*, which it readily adopts. As the state encourages corruption of the law, whose function has gone from serving and protecting to abetting crime and fueling destruction, only a perverse relation with speech can exist, or a ceaseless perversion of one's relation to speech.

In the absence of laws operating within the social order, censorship maintains the semblance of a barrier. In 1915, Sigmund Freud wrote: "Wherever the community suspends its reproach the suppression of evil desire also ceases, and men commit acts of cruelty, treachery, deception, and brutality, the very possibility of which would have been considered incompatible with their level of culture."[25]

Internal censorship (such as the moralization of religion) compromises the future of the subject and the social order. Under censorship, various degrees of perversion continue to operate in silence. In this context, the "community" doesn't allow the living to separate fantasy from taboo. ("I can wish to kill but it remains in a state of fantasy since, subject to law, the act is taboo and, in this case, I can only imagine it and fantasize about it.") The law of national reconciliation fuels hate among individuals and the public, and bars the possibility of conceiving of the Other (same and different) as a potential source of help and aid. Not knowing "who is who," the subject is constantly on guard and hunting for the suspect, in spite of, or because of, its inability to identify him, recognize him, or even ascertain his responsibility.

The only hope of "rehabilitation" lies in the possibility of giving a name, face, and story to this Other, even if he was a ruthless murderer. Anything less and the social bond remains a waking nightmare. Fear threatens social harmony and makes falling asleep, which is necessary for forgetting, impossible. In other terms, this law that authorizes the erasure of the Internal War produces just the opposite: *it makes it impossible to forget*. Representations of murder and violence are likely to be infinitely re-enacted in a myriad of forms in the social sphere. Indeed, erasing crime and acts of savagery ends up prolonging and accentuating them indefinitely, thereby extending, as Rachid Mimouni called it in his novel, "the curse" [*La Malédiction*].

When the state tries to make its practice of disappearance disappear

The way the political order treats History – and the struggle to consider it and analyze it as both a personal and collective responsibility – is a matter that extends beyond the situation in Algeria. To order the forgetting of murder, savagery, and the whole civil war to show some semblance of unity is tantamount to trying to erase the crimes and massacres committed in the name of God's law. Make no mistake. It isn't so much the Internal War that the law aims to erase from individuals' memory but, rather, the political implication of these crimes. Erasure is but a political move dealing strictly with political matters, things have come full circle, move along, nothing to see here ... However, erasure (even a well-orchestrated one) calls attention to the vanished history by its very blankness. It makes the illegible emerge. This is what I am calling *state terror*.

Between 1993 and 1996, there were between 15,000 and 20,000 disappearances in Algeria. For the most part, this consisted of men between 20 and 35 years old who were suspected without any evidence of engaging in "terrorism" and/or considered as likely to be "contaminated" by the Islamists.[26] Hooded officers would appear in the middle of the night at homes to make an arrest, telling families: "We're just going to question him and then let him go." Those who were picked up suffered a terrible fate: tortured to death or summarily executed. In either case, they never made it back home and their bodies were never returned to their families, nor were they given a dignified burial, which is to say, to be buried with a patronym. There were also countless neighborhood raids where the police swept up young men indiscriminately, most frequently in low-income neighborhoods in the city.

The *Collectif des familles de disparus en Algérie* ["Collective of Families of the Disappeared in Algeria"] notes that no list of names was ever established and that there are no institutions to turn to for families concerned about what happened to their children. Still today, mothers weep for their sons, wives for their husbands, fathers, or brothers. Mothers gather together in a plaza, just like the Mothers of the Plaza de Mayo in Buenos Aires, demanding that their children – or at least their children's bodies – be returned to them. The state claims that these disappearances were carried out by the Islamists, who also engaged in kidnapping, but they mainly kidnapped young girls and, to a much lesser extent, adult men. While the Islamists kidnapped young girls and women,[27] the state emptied houses of young men in a haphazard manner. This strange divide between sexes in the service of terror merits further investigation.

In 2015, the state considered the case of the disappeared closed, given that "the Charter of National Reconciliation" had granted families financial compensation provided they sign a death certificate. They were thus summoned to declare someone who had disappeared and whose body remained missing as dead. Demanding more information and refusing to sign the death certificate are ways of refusing to let "the practice of disappearance disappear." Families were implicitly called on to accept and consent to state terror. The judiciary was carefully kept out of this process. Numerous families refused to obey this law of *omertà* – and rightly so. The practice of forced disappearances was an instrument of state terror wielded against innocent people, everyday citizens guilty of no more than being in the wrong place at the wrong time. Still today, there has been no independent investigation to determine what happened.

What's more, the quasi-systematic practice of torture by police forces, while the bodies of countless Algerians – men and women – bear the marks of abuse suffered under the French military, merits further scrutiny. And what can be said about the repetition of torture among generations, with fathers and mothers tortured during the War of Liberation and sons and daughters during the Internal War?

The law of "national reconciliation" works to erase state terror and build in, and through, silence a level of consensus. Which is precisely what the mothers of the disappeared refuse to agree to by taking to the city streets. They are there to remind everyone of the disappearance of their children's bodies. In its deliberate absence of justice and acknowledgment, the state positions itself as the owner of bodies rather than the protector of them as citizens. Once again the argument can be made that terror operates by seizing real bodies. If Islamism, by hacking bodies to pieces and putting them on display, sought to go beyond death and to de-humanize and de-sacralize the body, it seems state terror, for its part, also operates by "killing the dead" and asks families to consent to this double murder (of the individual and of death) by providing them with financial compensation.[28]

The state's failure to acknowledge its responsibility and its desire to erase its crimes can only give the disappeared a terrifying power. In the absence of a defined limit (body, tomb), nothing holds back its ghost. It is nowhere and therefore everywhere. Its cries are heard by the living as an endlessly rustling noise as it refuses silences, a point to which I will return in chapter 9. How can we understand this strange matter where the state, meant to protect its citizens and enforce fundamental taboos, tries instead to make its practice of disappearance disappear under the pretense of treating a matter of "domestic security"?

The law of national reconciliation is thus paradoxically the institutional arm responsible for erasing the practice of disappearance. In the meantime, the Islamists have been able to take advantage of the legal opportunity they were given to be absolved of any responsibility for crimes committed. All this so that the state's responsibility is never exposed. How can we get out of this dialectic between a state of terror and state terror while the souls of the disappeared – with no bodies or names – wander by the thousands? What form of subjugation does this situation create for the living?

To press on with the matter of state terror, let's recall that, between April 2001 and April 2002, almost 130 people were killed and 500 more wounded as a result of a strong crackdown – once again – on the Kabyle population. It is worth noting that the so-called "law of national reconciliation" was voted for three years later, confirming the fact that this law essentially sought impunity for crimes committed by the state. On April 8, 2001, an 18-year-old high-school student with a typical Amazigh first name – Massinissa, who was a Numidian king – was killed on the premises of the national police. This sad episode led to a number of demonstrations which were violently repressed. Throughout the course of the year, there were many very violent altercations between the Kabyle population and the police. The protesters were demanding respect for human life and acknowledgment of their place as citizens as well as acceptance of both their "national" belonging and their cultural differences.

To these legitimate demands made by citizens the state responded with even more abuse of its power and disregard for the law. And this comes after a long history of repression and humiliation of the Kabyle population. The Amazigh demand for having its language officially recognized undermined the ideal of a pure Arabic and, for this reason, it was never given any serious consideration. It should be noted in passing that a template of the land's languages is found in the speakers of the Berber, Kabyle, and other languages (Chaoui, Touaregs, M'zab), who are living testaments of the land's plural and hybrid history. Note, too, that the suppression of Amazigh references goes hand in hand with the desire to systematically erase history and memories in favor of an Arabo-Muslim colonial history. The fighting between the state and the Amazigh populations confirms that internal difference is precluded from Algerian politics. Difference is considered not as an asset, but as a threat to a unified "Arab" identity.

The political order is struggling to move away from colonialism. In particular, it is struggling to deal with the impact of French colonization without substituting it with another form of colonization. Does this suggest that colonial occupations provoke a sort of fascination that becomes impossible to shake off? As we will see in the next chapter, the colonial occupation of land goes hand in hand with political impunity, with fratricide as a major consequence. But is this explanation good enough? Is this understanding of fratricidal pleasure yet another way of displacing – and therefore neglecting – the responsibility of those involved?

8
Legitimacy, Fratricide, and Power

What eternal curse condemned them to discord?
Rachid Mimouni, 1993[1]

[Y]our country has been offered for sale. ... [U]nless the guilty are punished, what will remain except to pass our lives in submission to those who are guilty of these acts? For to do with impunity whatever one fancies is to be king.
Sallust, 40 BCE[2]

Countries are plagued – implicitly or explicitly – by the question of impunity for criminals. Under the surface of the political order rages a battle between promoting an open society founded on social harmony and returning to a closed society fueled by hate and murder.

The consequences of impunity are at the heart of the story of Jugurtha, King of Numidia (present-day Algeria), who lived from 160 BCE to 104 BCE. This king, who became the pride of the Amazigh people and therefore all Algerians as an unparalleled warrior who had fought against Roman invaders, has become a living legend. His status as hero was reinforced during French colonization and the War of Liberation. The legend, as retold by countless stories, supports Freud's argument concerning how reconstruction is at work within the subject and people's imagination, for the beloved tale of the legendary hero masks a very different story, one centered on fratricidal pleasure.

Freud had identified the lie inherent in the hero's creation, a hero who supposedly accomplished his feat single-handedly. In other words, the hero seizes power, and in so doing he removes two groups: the enemy and the brothers with whom he carried out the murder. The question of power is central to the formation of the hero figure. It is worth examining in detail the repressed site where he comes into power by eliminating his brothers. Attributing his heroic act to him alone thus distracts from a murder and the elimination of those who allowed him to acquire his power. As we are nearing the close of this work, my argument should be coming into clearer focus: the political order is founded on an endless conflict between liberty and terror, openness and closure, and, ultimately, fraternity and fratricide. No matter how the political (and subjective) order is constructed, it is always at risk of swinging abruptly from one side to the other. Freedom of speech and fundamental taboos are never established once and for all; rather they are subject to *reversals*, *refusals*, and *détournements*. Our work to establish them is therefore never complete.

Today, the wars of conquest have hardly disappeared. They continue to rage behind other signifiers. Thus, questions concerning the fate of unpunished crimes apply equally to tyrannical leaders and to so-called "democratic" regimes. The Roman historian Sallust speaks to this issue by citing a speech given by the Republican Memmius before the general assembly in Rome in the midst of war against Jugurtha:

> For which of you dared to refuse slavery? For my own part, although I consider it most shameful for a true man to suffer wrong without taking vengeance, yet I could willingly allow you to pardon those most criminal of men, since they are your fellow citizens, were it not that mercy would end in destruction. *For such is their insolence that they are not satisfied to have done evil with impunity, unless the opportunity for further wrong-doing be wrung from you; and you will be left in eternal anxiety, because of the consciousness that you must either submit to slavery or use force to maintain your freedom.*[3]

Jugurtha: a fratricidal hero

The legendary figure of Jugurtha has sparked an often-overlooked debate about unpunished crimes at the heart of republics. These crimes are overshadowed by political and economic interests, which, as real measures of power, fuel the desire of republics to conquer more land. Jugurtha, our fearless hero who took on the Roman Empire, is still admired and praised today. The image of his glorious power probably helps counter the deep-seated feeling of *hogra* instilled by the oppressor. Nevertheless, the beating heart of this hero's story was kept out of the wonderful legend and has quite simply been forgotten. Jugurtha was born, as they say, out of wedlock. His father, Mastanabal, of Greek and Punic origin, is the brother of Micipsa, King of Numidia. Both are the sons of the former King Massinissa, the emblematic figure of the land. Jugurtha is thus the nephew of Micipsa. Upon his birth, Jugurtha wasn't recognized as an heir to the royal family, since he was born from the union of his father and a concubine/slave, which is to say, a "free woman." This so-called "illegitimate" birth excludes him from his heritage: Jugurtha is to the royalty what Jean El Mouhoub Amrouche deemed the "bastard" (see chapter 4 above), a term he also uses to describe the experience of the "*indigène*" of Algeria, who is "deprived of past and future generations." And so, how does Jugurtha go from being excluded from royalty to King of Numidia? And where does his story of being an illegitimate child lead him?

Micipsa takes note of Jugurtha's talents as they are recounted to him from both inside and outside of Numidia. The King tells his "illegitimately born" nephew: "In Spain the name of our family has been given new life. Finally, by the glory you have won you have overcome envy, a most difficult feat for mortal man." This seemingly enigmatic statement becomes clearer once we recall that the King was made very uncomfortable by the strength of this young man, whom he considered deep down as a potential enemy to his kingdom. But as the General of Numidia heaps praise on Jugurtha while speaking to Micipsa, praising his qualities as a warrior and strategist, the King finds himself in a bind. He wants to send him away because he is a threat. However, by ousting him, he risks sparking a rebellion or even war. Micipsa develops a plan to turn his enemy into an ally. To accomplish this, he adopts him and recognizes him as a legitimate child, on the same level as his two sons, in order to thwart his own "desire" to eliminate him. Being adopted, however, does nothing to repair Jugurtha's devastating feeling of illegitimacy. Nor is this feeling appeased by his status as a powerful warrior.

Jugurtha isn't fooled by the outward display of affection from the King (who is also his paternal uncle). He knows his real feelings and is aware that being recognized as his son only masks the fear of an internal war. As he nears death, the King summons his inner circle to announce the news of Jugurtha's adoption. Addressing Jugurtha before this group, he says:

> When you were a small boy, Jugurtha, an orphan without prospects or means, I took you into the royal household, believing that because of my kindness you would love me as if you were my own child. And I was not mistaken; for, without speaking of your other great and noble actions of late, on your return from Numantia you have conferred honour upon me and my realm by your glory, and by your prowess have made the Romans still more friendly to Numidia than before.[4]

This is at a time when Rome, after its victory over Carthage, controlled a large part of the Mediterranean basin. The historian Houaria Kadra, a specialist of this period, informs us that Numidia supplied Rome with wheat and elephants, which were used in battle – let's recall that the supposed offense that triggered the French invasion of 1830 was also related to a debt the French had contracted toward Algeria for unpurchased wheat (see chapter 2 above). Micipsa placed Jugurtha first in line for the throne even though he had two younger sons: Hiempsal and Adherbal. Hiempsal was said to be jealous of his brother's success and contemptuous of him, reminding him at every occasion of his status as a child "born out of wedlock" and therefore of his illegitimacy. From the very first meeting between the brothers to discuss matters of governance after the King's death, the feeling of offense returns: "Hiempsal," Sallust recounts,

> who was naturally haughty and even before this had shown contempt for Jugurtha's inferior birth because he was not his equal on the maternal side, sat down on the right of Adherbal, in order to prevent Jugurtha from taking his place between the two, a position which is regarded as an honour among the Numidians. ... Hiempsal reminds Jugurtha that he was given his place in the kingdom through adoption, and, for this reason, *he isn't a legitimate child of the kingdom*. This remark sank more deeply into Jugurtha's mind than anyone would have supposed. He was deeply *hurt, offended and humiliated* ... . And so, from that moment on,

> he was prey to resentment and fear, planned and schemed, and thought of nothing except some means by which he might outwit and ensnare Hiemsal.[5]

Here, too, *offense*, *humiliation*, and *contempt* – signifiers all embodied by the polysemic, Algerian-Arabic term *hogra* – are at play. Although they all work on a different level, these three states indicate an existence overcome by pain and constitute a form of internal murder. It is a logical consequence that one responds with violence. It would be interesting to examine what role offense has played in every war. Note that offense and "offensive" (a war term) come from the same root. When offense is felt and experienced, it seems to give rise to an almost irremediable offensive attack.

The legend of the French conquest contains at least two offenses: humiliating remarks about Islam allegedly made by French representatives; and the offensive response of the Bey, who slapped the representatives with his fan. This offense presented an opportunity for conquest. France attacked the port of Sidi-Ferruch in the bay of Algiers. The question of offense is at the heart of this almost legendary story. Except that the real offense remains repressed and hidden. The real offense for France was the defeat handed to it by Great Britain and the European coalition in 1815. The war between these two countries continued to be fought as a competition of colonial expansion, which only fueled the rivalry between the two empires. Colonized territories thereby became the displaced battlefield – without suffering any real battle – of an ongoing war between the two powers.

The offended Jugurtha seeks revenge for this comment on his illegitimacy by waging a bloody fratricidal war, before waging another war against Rome. One war is intimately tied to the other. The war for freedom against the foreigner came in the wake of a fratricidal war, which has been carefully left out of the legends, myths, and hero narratives. It is hard to ignore the fact that the internal – fratricidal – war and the war waged against a foreign occupant were fought on the same land at the same time. Did this set the template for the War of Liberation, a template undisturbed by the passing of time? Are we condemned to an "eternal curse" (Rachid Mimouni)?

Shortly after this episode of offense (contempt/humiliation), Jugurtha kills his brother Hiempsal and unleashes a merciless war on Adherbal. After the death of Hiempsal, Numidia finds itself plunged in terror. The land becomes divided in two, with either part supporting one of the two brothers (Jugurtha and Adherbal). A war between two brothers ensues, with only one brother considered legitimate. Adherbal summons the Senate to Rome so that it may intervene in a fratricidal battle he finds deeply unjust. Jugurtha is very skilled in the art of corruption and *détournement*. He appeals to the Senate and the political figures of the Roman Republic with massive quantities of gold and wheat, with the hope that they'll look past his crimes, in the spirit of Republicanism!

Full-on anarchy breaks out, with no end in sight for the fighting. Adherbal proves no match for the all-powerful Jugurtha. Jugurtha, determined to be the sole successor of the kingdom, is hardly satisfied by Rome's proposal to divide the land between the two brothers. Once again, the subject beset by a feeling of illegitimacy is driven by "envy" and a mad desire to possess everything and to rule single-handedly. Sallust explains that the Romans were seduced by Jugurtha's gifts. Easily giving in to corruption, they withhold from judging his criminal acts and effectively grant him impunity. Despite Adherbal's calls for help as Jugurtha hunts him down, Rome remains the silent and passive witness of this war between brothers. Adherbal gives a moving testimony before the Senate. He doesn't hold back when addressing, once again, the matter of birth, which is to say, his blood rights, and labels Jugurtha an "impostor." He believes Jugurtha is a usurper who is acting against his father's own word. Once again there is a shift from internal enemy to external enemy. Between the two wars, can the so-called "legitimate" war (against a foreign enemy) temporarily hide the enemy from within (the brother)?

Here is Adherbal addressing the Senate of Rome:

> If I had no other reason for asking the favour than my pitiable lot – of a late king, mighty in family, fame and fortune; now broken by woes, destitute and appealing to others for help – it would nevertheless be becoming to the majesty of the Roman people to defend me against wrong and not to allow any man's power to grow great through crime. ... Woe's me! O my father Micipsa, has this been the effect of your kindness, that the man whom you put on equality with your own children, whom you made a partner in your kingdom, should of all men be the destroyer of your house? Shall my family then never find rest? Shall we always dwell amid bloodshed, arms and exile? While the Carthaginians were unconquered, we naturally suffered all kinds of hardship; the enemy were upon our flank ... .

> After Africa had been freed from that pestilence, we enjoyed the delights of peace.[6]

Jugurtha, unpunished for his first crime, seeks to kill again, this time by murdering his other brother Adherbal, with the passive complicity of the Roman Senate. Rome prefers to turn a blind eye to the crime and proposes a compromise between the two brothers rather than risk losing its advantages in wheat, gold, silver, and war animals. Jugurtha, heedless of the laws of the Republic, pursues a relentless war against his adopted brother: pillaging, killing, laying waste to cities and villages. His feeling of illegitimacy pushes him to seek absolute power at all cost: he wishes to be the only child. He becomes a criminal and has Adherbal tortured to death. The historian Houaria Kadra, who has commented extensively on the wars of Jugurtha, recounts that the Romans also had taken advantage of the various fratricidal rivalries between Jugurtha and Hiempsal/Adherbal, and between Jugurtha and another paternal half-brother, Gauda.

The Roman Republic views Jugurtha's second crime as an offense, as Kadra explains: "Rome had to avenge the affront it suffered."[7] And so Rome declares a ferocious war on Jugurtha that will last seven years – almost the same length of time as the Algerian War of Liberation. Kadra recounts these barbaric wars in vivid detail. Her careful account, which supplies a trove of information that is missing from Sallust's telling, has informed my own reading here.

Jugurtha, pulling off one exploit after another, continues to wage devastating battles against Roman troops. He remains elusive in spite of Rome's aggressive pursuit of him. The Romans are forced to use all possible means to capture him, employing a whole array of tactics. Jugurtha's principal offense here is thwarting and making a mockery of the Empire's army. The political heads of the Roman Republic are divided between those who are unbothered by corruption and wish to protect him and those who are taking a stand against the violation of the laws of the Republic. They call on Bacchus, King of Mauritania and Jugurtha's father-in-law, to trap his son-in-law. This is a full-on family war, at once fratricidal and almost filicidal. In return, the Romans pledge to expand Bacchus' reign, granting him both Numidia and Moorish country. For seven years, Jugurtha remains undefeated and continues to taunt Roman power. But his father-in-law, Bacchus, who had pledged to protect him, betrays him and delivers him to the Romans. Gauda, his half-brother, is granted power to rule over this land after striking a deal with Rome.

Unpunished crimes within the Republic

Jugurtha, a "free man," repeatedly insists that "he's not Roman." This is what the Algerian legend has retained from this story: a hero who refuses to be subjugated and who defines himself in strict opposition to the Other. But this positioning by opposition is but another form of subjugation that prevents subjects from finding their own defining features. They remain, in fact, trapped in a self defined by the Other (a self that is not Roman). The field of indeterminacy finds itself once again closed off as the self is caught in a binary logic. The Republican Memmius' question cited above regarding the consequences of failing to punish Jugurtha for his first fratricidal murder is in this respect enlightening: "For which of you dared to refuse slavery?" He later elaborates on this thought: "[Y]ou should not insist upon ruining the good by pardoning the wicked. Moreover, in a republic it is far better to forget a kindness than an injury. The good man merely becomes less active in well doing when you neglect him, but the bad man grows more wicked."[8]

The exploits of this legendary heroic figure – an illegitimate child – mask the fratricidal drama orchestrated in collusion with the Republic. A detail stands out in his story: our illegitimate hero has become a master in the art of the corruption and the *détournement* of laws, institutions, and finally the political order. It seems almost as though this were set in present-day Algeria.[9]

Jugurtha and his sons are captured and taken to Rome, where they suffer terrible humiliation. Jugurtha is tortured and imprisoned in a cave, where he is starved for having offended the Romans. Six days later, he is dead. The consequences of this episode are of even more significance. The Roman Republic suffers its first civil war in 88 BCE right on the heels of its war against Jugurtha. According to Kadra, two figures, Sylla and Maurius, fight to claim credit for Jugurtha's capture. The civil war threatens the Republic, as both Sylla and Maurius claim themselves the hero of the war against Jugurtha. But History ignores remembrance. A war against the Gauls is looming, the very people whose descendants will later take over the land controlled by Rome.

How strange this history of a land and people where the past, which can never seem to pass, is perpetually resurfacing, and yet remains out of reach of memory. Memory of the offense and of the war are seemingly embedded in the land, giving rise to History, but a History that appears timeless. From war to war, the eternally unremembered seeks to become a part of remembering but is flatly and repeatedly denied. War between the Romans and the Gauls is inevitable. As Sallust makes clear: "[T]error at this had made all Italy tremble. The Romans of that time and even down to our own day believed that all else was easy for their valour, but that with the Gauls they fought *for life and not for glory*."[10]

Violent fratricidal fighting is often triggered by the occupation and control of land. Yet this holds true only if fratricide is considered the eternally unremembered part of the political order. Again, civil wars spring from the resurgence of what the political order has repressed. From fraternity to fratricide, there is a particular kind of political memory where forgetting and the *inability to forget* co-exist. In some cases, repression is predominant (as in democratic societies) and produces forgetting, but, in others, this mechanism is "faulty", allowing the *inability to forget* fratricide to surge forth.

Republican fraternity is a reversal of fratricide, which constitutes the political order. Jacques Lacan addressed this in 1970:

> The energy that we put into all being brothers proves that we are not brothers. Even with our brother by birth nothing proves that we are his brother – we can have a completely opposite batch of chromosomes. This pursuit of brotherhood, without counting the rest, liberty and equality, is something that is pretty extraordinary, and it is appropriate to realize what it covers.
>
> I know only one single origin of brotherhood – I mean human, always humus brotherhood – segregation. We are of course in a period where segregation, ugh! There is no longer any segregation anywhere, it's unheard of when you read the newspapers. It's just that in society – I don't want to call it "human" because I use terms sparingly, I am careful about what I say, I am not a man of the left, I observe – everything that exists, and brotherhood first and foremost, is founded on segregation.[11]

Segregation creates subjects with an exceptional status, which is to say, outside of the common laws founded on fundamental taboos. This state allows fraternity to rediscover its original impulse to kill. Under colonialism, segregation, mass murder, and censorship were foundational in the formation of the social bond. They waved the banner of universalism. As such, they constitute what is *impossible to forget* about the political (and subjective) order in its present manifestation.

The legitimacy the French conquest claimed for itself

My aim in this work has been to translate specific erasures carried out under colonial rule into legible script. In this process of reconstructing a historical background, it has become clear that coloniality is propped up by a whole series of myths and legends, designed to console orphaned and abandoned sons left out of history. The immigrants, "the lost children of Christian culture,"[12] arrived as heroes to conquer the other "savages," the now illegitimate sons. The first colonizers during the period of conquest were clearly the "rejects" of a young French Republic. As Lamartine expressed it in 1834: "Dear sirs, there you have colonization! It doesn't create wealth right away, but it creates a mobile workforce; it multiplies life, encourages social movement; *it preserves the body politic both from depleting its force and from an overabundance of unemployed energy, which sooner or later explodes in revolution or disaster.*"[13]

The colonial order's belief in the existence of a land without history or culture produces in fact the blank space it wishes to see. This belief masked another reality about the future French Republic: the fear provoked by the threat of a grueling political battle between monarchists and republicans waged on its own land, a real possibility that had to be warded off like a curse. The specter of civil war had also loomed over Rome after Jugurtha was captured and put to death. The Roman Republic thus went from one war (against Jugurtha) to another (civil war). Several centuries later, Gallic descendants took back their agricultural heartland from their vanquished enemies. From one war to another. The idea of land as blank space allowed the French victors to ignore the longstanding war fought over land (Roman/Gauls).

Coloniality is founded on a compelling origin myth for both of the land's parties. In reality, conquest is but the continuation of a history that has been unfolding over a long period. France justified the legitimacy of its invasion by citing Algeria's *Latin past*, whereas the real motivating force came from its rivalry with Great Britain.

It is no surprise, then, that post-Independence Algeria later drew on the Arab conquest in its effort to escape French domination and assert its own legitimacy.

The French conquest claimed the land's Roman past as its own. It was a matter of taking the place of the victors (the Romans) and transferring the offense they felt from Great Britain to the "*indigènes*." What role, then, does the imaginary play in conquests and the seizing of territory? The historian Todd Shepard argues that there is a clear continuity between the French conquest and the land's Roman history. The terms he uses show, once again, that the imaginary is constructed around notions of succession and legitimacy for France. Algeria, in this case, was far from being a land without History. On the contrary, this land was brimming with memories blanked out in order to legitimize taking possession of the land, its history, and the spirits of its inhabitants. Algeria was a spatial archive commemorating other battles of European conquest. Thus, the unremembered is historically determined, time and time again. The violence of History is woven into the very fabric of this political moment.

There is no clearing on the horizon:

> Only three years after its occupation of Algiers, the French army invited archaeologists to visit the land and, in 1837, it established a commission to explore Algeria's archaeological resources. At the end of the 1880s, the newly created French School of Rome encouraged all the archaeologists and antiquarians it hosted to do research in Algeria. All of this accumulated knowledge gave credence to the comparisons made by the military, political leaders, and a growing number of scholars between ancient Rome's history in North Africa and France's effort to take control of the same space. The French saw their own role as bringing order and civilization to North Africa in parallel to ancient Rome's. This allowed them to view their effort as bringing order back to any area destroyed by Arab and Muslim invasions. This idea was directly expressed by countless classical scholars. A telling example of scholarly exaggeration among others was the claim that the land of Algeria had been the granary of wheat for the Roman Empire, which led to the belief that the Ottomans and the indigenous populations had destroyed this heritage.[14]

Unremembering is like a hyperactive form of memory that haunts a land's history and geography (see chapter 6 above). A case in point is the repressed memory of Mitidja, a key battleground during the

War of Liberation and a site of terror during the Internal War. Is what lies out of reach of memory for the subject and the political order found in the outskirts of History, embedded in the land?

Colonial literature from this period created the myth of a triumphant France restoring the glory of the Roman Empire, which was laid to waste by the Arab invasion. The writer Louis Bertrand (1866–1941) is especially enlightening on this point, showing how this belief underlined much of the work and novels of the period. The Gauls stopped fighting the Romans. Their descendants came to rescue them from the Arabo-Islamic catastrophe in order to pursue, through conquests existing on the level of the real and the imaginary, its rivalry with Great Britain. This myth anchors the French colony in its Latin and European history. A powerful myth determines in no small way the course of History.

Algeria becomes a space, then, imbued with the spirit of Europe. In his "Short Guide to Towns without a Past," Albert Camus writes:

> The softness of Algiers is rather Italian. The cruel glare of Oran has something Spanish about it. Constantine, perched high on a rock above the Rummel Gorges, is reminiscent of Toledo. But Spain and Italy overflow with memories, with works of art and exemplary ruins. ... The cities I speak of, on the other hand, are towns without a past. Thus they are without tenderness or abandon. ... These towns give nothing to the mind and everything to the passions.[15]

Algerian cities, according to Camus, are devoid of memory and lack their "own" history. They are blank spaces ready to be inscribed by reminiscences of an elsewhere: Europe is vividly present. These cities become European in the conqueror's near-hallucinatory vision. They willingly embrace the language, history, and geography of the colonizers, who become the land's "first inhabitants." The "*indigène*" simply disappears with this logic of negation. He or she is the site's unwelcomed Other.

This erasure allows "French Africa" to act unquestioningly as a natural extension of "Latin Africa" (Bertrand). For Bertrand, in Algeria, Rome can be seen everywhere. Coloniality has no beginning; from the Roman Empire to the French Empire, the land has always been Latin. This isn't a matter of appropriation or expropriation of land, but a legitimate continuity of culture, religion, and originary language. The Latin world was always already there, waiting to be reoccupied by its rightful owners. This point is essential for understanding how coloniality, *for both the colonizers and the colonized*, springs from an origin myth that implies both recovery (for the conquerors) and loss (for the "*indigènes*," who, for their part, will succumb to this same fiction after Independence). An ever-lasting "Latinity" as a figment of the imagination that conveys both birth and death, expressed via the land. "Thus," writes Louis Bertrand,

> was maintained the age-old tradition which makes Africa the tributary of Empire. But more than anything else, the Roman ruins that surround it like an unbroken network perpetuated the obscure memory of this tradition. The seal of Rome remained visible everywhere. This is why when the French, Spanish, and Italians wasted no time settling there after 1830, they had the illusion of returning to their abandoned domain and of taking back their property. There is every reason to believe *they were seizing these properties for good.*[16]

The land is so haunted by its Latin past that it makes everyone believe the forced disappearance of the autochthonous ("*indigène*") population has merit. The foreigners on this land that once belonged to Rome must go. This is the thinking behind the settler's colony, whereby the position between the colonizers (the conquerors) and the "*indigènes*" has been flipped, making the latter a minority population. As the historian Olivier Le Cour Grandmaison explains:

> By their presence alone, the colonizers built up a relation of power that is all the more favorable to them as the "Arab factor," knowing its irreversible situation and with no hope of recovering its land, "becomes more and more isolated and little by little dissolves," as Tocqueville, who had witnessed this in and around Algiers, said. He was convinced that the Muslim population would continue to decrease, while the Christian population would only increase. … Clearly, Tocqueville remembers what he learned in the United States.[17]

The colony's double dealing of birth and death, though given distinct expressions, is consistently the same wherever it is established: rebirth for some, disappearance and mass murder for others. In fact, this binary logic defines it. The erasure of this colonial ideology serves the rewriting of a mythic history, one that casts itself as the official narrative on which the political order is founded. Supplying the missing link of Latinity reframes the history of

France's conquest as a legitimate undertaking. The "sons" rejected by their French homeland reinvent their longed-for legitimacy and are given a second chance in this conquered land.

The myth at the center of coloniality clearly inverts the categories of familiar and foreign. Labeled foreign and undermining France's vision of reunion, the autochthonous population is to be removed from the picture. It must disappear to allow the emergence of an unmistakably familiar world, an isolated community built around Latinity. The notion of purity within coloniality aims to construct a white ("whitewashed") European unity. It is no surprise, then, that Algiers is named "*Alger la blanche*," or "whitewashed Algiers," by the colonizers, an expression that is still used today.

The origin myth perpetuates the notion of a race whose legacy must be preserved and continued. The sexualization of relations[18] through the creation of a dominant and a dominated class (tied to the notion of possession) and through the signifiers and practices employed under colonialism and during the war gives this Latin origin its sexual dimension: widespread rape of women and young girls, men emasculated and bodies mutilated constantly, rendering men impotent through torture (electric shock), and so on. It was a matter, on the one hand, of increasing the fertility of the Roman Empire and, on the other, of reproducing "Europeans" through a policy deemed hygienic. Let's recall that "European" is the historical term given to the colonizers.

The Latin Empire is the primal scene for the French Empire. All-conquering France is bound up with a split partner that fuels its own imaginary: at times Latin (to be rescued, loved, protected, glorified) and at others "*indigène*" (to be hated, beaten, destroyed, made to disappear). We are at the heart of the mythic, impassioned scene of coloniality.

The impassioned scene of coloniality

To those lost in their own fantasy, the traces of a Roman past marking this land appear brimming with life; that particular past seems to them much more real than the inhabitants reduced to the "shadows" of a dying past, according to Louis Bertrand. The relation between the visible and the invisible, the living and the dead, the past and the present has been flipped. The hallucinated reality allows the visible traces to take shape and become a living, ubiquitous force, and therefore allows for a reunion with the lost object (Latinity) to be imagined.

From this perspective, the feeling that Latinity had been lost or damaged turns into a feeling of rediscovery and redemption within French coloniality. This is the mission to rescue the Latin Empire. Bertrand gives us a glimpse into the madness of recovering an object from the past, one that was believed to be lost forever and which has reappeared in all its splendor. There is a slight shift that takes place from fantasy to an ideology of purification. Referring to the vestiges of Roman society in Algeria, Bertrand writes:

> The imprint was such that, to trained eyes, the vestiges can still be seen in the customs of the *indigènes*. Moreover, the Arab conquerors added nothing to the heritage of Rome; instead, they set about destroying everything that was not imposed on them by force of habit or climate. After razing everything to the ground, they didn't know how to build things back up; and so, this conquered country, where they stayed like campers on the move, is for today's tourist like a vast museum where everything has remained intact, since the day when the temples and the *arcs de triomphe* designed by architects erected their solitary walls in the midst of burned-down cities. This is why these ruins are *more striking* than anywhere else. Their *evocation of antiquity* is not affected by the proximity of modern buildings, and even the poverty-stricken beings who, wrapped in their white-wool coats, wander around them seem to be *shadows emanating from the same time period of the ruins*, so little has changed in their habits and attire, while everything was changed in the Latin world. They look like *millennia-old witnesses* of the catastrophe that suddenly abolished the abundance and life of this once industrious and *fertile* Africa.[19]

This strikingly haunting and erotic passage reveals the fantasy-driven project of colonization and its constant sexualization. It is a question of increasing the fertility of the Roman Empire and making the current inhabitants shadows, witnesses, and the remains of a History with no past and no future. The "*mission civilisatrice*" aims to revive the vestige-object, impregnating it with the Latin seed so that it comes to life: the colonized land becomes clearly divided between an object glowing with desire (Latinity) and pure waste to get rid of (the "*indigène*").

The story of Jugurtha allows us to consider how the return of the feeling of illegitimacy impacted the "*indigène*." Namely, it constituted a terrible offense which, in turn, called for a fratricidal offensive. The origin myth of coloniality thus gives the conquest of Algeria its historical legitimacy. The cycle of killing, disappearance, and annihilation is sustained by a desperate search for legitimacy, to which there is no end in sight. The erasure of the colonial *cause* is a tool used to rewrite in bold letters a text that casts itself as new whereas it is simply part of the blank space left by a disappeared text: "We must hope," says Bertrand, "that Africa will once again become the great intellectual garden where East and West sowed the seeds of their religions and knowledge. The Latin spirit would undoubtedly find there an opportunity for renewal and rejuvenation."[20]

The truth doesn't lie in Algeria's supposed virgin land – with its sexual connotations – but in the signifiers of language and history. Indeed, "he who is not Roman" (the literal meaning of the word *Amazigh*) becomes in the imaginary an enemy of Latinity. He is thus to be removed from the ancient city so that the city can regain its youth and vigor with no obstructions. The Amazigh (he who is not Roman/a free man) would also later become the "foreigner" or "stranger" to the land. He embodies the ancient enemy, the Romans, who in people's minds were seen as allied with the French conquest after the fact (and this foreigner, a genuine member of the autochthonous population, continues today to be fought by the Algerian Republic, as can be seen in its years-long repression of the Amazigh people).

The obsession with Roman ruins was poignantly described by Albert Camus in his essay "The Wind at Djemila" (the site of Roman ruins):

> *There are places where the mind dies so that a truth which is its very denial may be born.* When I went to Djemila, there was wind and sun, but that is another story. What must be said first of all is that a heavy, unbroken silence reigned there – something like a perfectly balanced pair of scales. … In the great confusion of wind and sun that mixes light into the ruins, in the silence and solitude of this dead city, something is forged that gives man the measure of his identity. It takes a long time to get to Djemila. *It is not a town where you stop and then move further on. It leads nowhere and is a gateway to no other country. It is a place from which travelers return.*"[21]

Following Bertrand's lead, Camus calls attention to a consuming obsession with the land's supposed Latinity, which attests to a rare combination of an excess and lack of remembrance. Colonial literature thus reveals the ideology of coloniality and shows how "Algeria" is the space of an already existing Latinity, a land awaiting occupation. The elimination of the "*indigène*" becomes in this context a political necessity to allow a spectral memory to flourish, a memory that is at once Latin and blank.

The specter of discord: *el Fitna*

The construction of collective memory is complex, caught between forgetting (remembering) and the inability to forget (a hauntingly real and forceful presence). Individuals are subjugated to History through this tension of collective memory. After examining the re-enactment[22] of fratricide in battles over legitimacy, I will now turn to how it is involved in the spread of Islam.

Discord among brothers forever haunts the minds of Muslims. The question of a legitimate successor has occupied a major place in Islam since the time of the Prophet Muhammad. The obsession with purity among believers is a direct response to this fundamental illegitimacy. Indeed, the hard line taken against divergent opinions, varying interpretations, and inherent contradictions aims to maintain the illusion of purity. The question of legitimate successor has far too often caused fratricidal wars within Islam. According to the Tunisian scholar Héla Ouardi, near the end of his existence, many assassination attempts were plotted against the Prophet by his brothers, who coveted his place of unique and legitimate successor.[23] His closest companions devised schemes to prevent him from naming his cousin and son-in-law Ali as his successor. Let's recall that the separation between Sunni and Shiite Islam is tied to a bloody war between "religious brothers" (literal translation of the Arabic *Ikhouas fi el din*) over the question of legitimacy – a fight that is still raging today. The question of designating a successor deeply troubled the Prophet at the end of his life: "The warning signs of discord and fratricidal fighting that have been tearing Muslims apart for centuries were already present when Muhammad was on his death bed."[24] Fratricidal power structures the relations between a group of brothers who remain isolated from a maternal or female presence. The exclusion of women in this system is a necessity and a way of maintaining a pact of pleasure between men. In other words, Islam's form of fraternity sought between men masks the fratricidal struggle affecting its own power structure, a structure that also stokes a disdain of women.

As Muhammad feels his end drawing near, he decides to name Ali as his successor. This causes his closest companions, Abû Bakr and 'Umar, to launch a series of conspiracies with the hope of changing his mind. Warned just in time by the angel Gabriel, the Prophet narrowly escapes assassination on two separate occasions. Ouardi informs us that Ibrahim, his only son, born out of his union with Mariâ, died as an infant (twenty months old). Muhammad is overcome with sorrow. Gazing upon a mountain, he states: "Oh mountain, if you had to bear the sorrow that I bear, you would have collapsed."[25] This test of sorrow stemming from the loss of his only child reinforces his concerns about finding the right successor. Muhammad names Ali by divine order. However, despite the absolute power this order holds for believers, Abû Bakr and 'Umar reject his decision. The Shiites who embrace Ali are named the "refusers" in Arabic (*el rawafid*): they refuse to recognize the Prophet's closest companions, while the latter group wishes not to recognize Ali as a legitimate successor.

Each of the Prophet's two wives is the daughter of one of his companions (Hafas is the daughter of 'Umar, 'Aisha that of Abû Bakr). They play the role of messengers for their respective fathers, informing them of all of Muhammad's secret decisions. Succession is a family matter, made up of a series of alliances and betrayals. 'Aisha strives to grant her father the place of rightful successor, which has been her fantasy since she was a little girl. For her, only her father should take the place of her husband, the Prophet. A fight for this position ensues between the Prophet's father-in-law and companion (Abû Bakr) and his son-in-law (Ali).

The Prophet's word designating Ali as his legitimate successor is disavowed by his fathers-in-law, who are his traveling companions. The divine order is thus not recognized by them. Matters only get worse for the Prophet when others find out he is unsure about his paternal lineage. The question of Muhammad's father is indeed an enigma. Muhammad was thought to have been born two to four years *after* his father's death. The person he understood to be his father therefore couldn't have been his biological father. This enigma has puzzled theologians and historians of Islam. The legend concerning the Prophet's confused paternal lineage evokes the fabulous myth of the "sleeping child" common to many Muslim Amazigh societies: *el ragued*.[26] The story (which is still told today) has it that women can bear a child for much longer than nine months. The length of the pregnancy can be indeterminate and may depend on divine circumstances rather than on biology. This myth has given rise to countless debates between theologians concerned with the question "Who is the father?" as the child can remain in the mother's womb for up to four years. This myth also has it that some newborns can be born with a full head of hair and teeth.

The question of paternity is in any case a source of trouble for the Prophet. He is plagued by the question of legitimacy on two fronts, as both father and son: "Muhammad himself is worried about his origins. A hadith has him insisting on the nobility and purity of his origins, stating unambiguously that his parents conceived him in wedlock and not in an extra-marital affair."[27] For, in reality, the Prophet's father may have been a slave, and the myth of the sleeping child would serve to hide this origin. Like Jugurtha, this would trouble the Prophet Muhammad's relation to power, forcing him to grapple with an overwhelming feeling of illegitimacy. It bears recalling in passing that the Algerian political order has wavered since its foundation between historical and religious legitimacy in a context where the father's governing function has all but disappeared. This is why the war over legitimacy is far from over: from Jugurtha to Muhammad, the question of the legitimacy of power has only misguided and blinded a political order that has forever harbored a hidden family history and an experience of illegitimacy.

From this perspective, Muhammad was seemingly prevented by his companions from writing the testament where he was to name Ali as his successor. He was forbidden from making this pronouncement. Ouardi recounts how Muhammad had prophesied this moment: "I will put my wishes in writing, for fear that envy may flare up and some may say 'This belongs to me.'" Elsewhere, he says: "You will be humiliated after me."[28] The pronouncement of succession was never written, which may explain the endless war to establish absolute truths. Let's recall that a number of Islamists waged an unrelenting war against the pen, replacing ink spilled with blood. What a strange and spiteful parallel this offers with the unwritten testament. How can we understand the incessant trouble of discerning a transmissible trace? Or, recalling Rachid Mimouni's question: "What eternal curse have we been condemned to?"

Finding pleasure in fratricide is a frequent reoccurrence in history that eludes remembrance. In other words, it is left out of the historical record and yet it is its driving force. Brothers killing brothers is re-enacted without leaving any traces on which an account of it could be based. Without this account, there is no reckoning, and therefore no end to it. Once again, fratricide is lost behind the praise of martyrs (see chapter 4 above). The discord sown between brothers, *el Fitna*, looms over the account of a death caused by the Prophet's closest companions, and has remained unresolved.[29] The praise of martyrs is inseparable from fratricide in the Islamic tradition. The division between Sunni and Shia Muslims is but a consequence of this. The caliphate (empire) clearly arises from this history and perpetuates it. In Algeria, the *Ikouas mouslimounes* (Muslim brotherhood) killed each other to become the (only) sons/FIS. The formation of this political party (FIS) was seen as being legitimized by the state, and its dissolution could only mean the resurgence of killing. This unforgettable moment in History and its consequences show how the hopeless search for legitimacy at the heart of the political order repeatedly ends in catastrophe. Is legitimacy only reserved for the martyr, since he bypasses the father altogether by addressing God?

For Héla Ouardi, the assassination of the Prophet is a case of parricide. In light of her own research, I'd argue rather that it is a case of fratricidal murder, even though it strives to conceal this fact. Muhammad was kept from assuming the place of father by being prohibited from writing a testament. He was robbed of the right to leave an inheritance, leaving his word and memory abandoned. The fratricidal murder that took place is kept out of the narrative and memory of Islam. Consequently, discord cannot be contained and explodes in a series of acts. As such, inheritance has been a devastating problem for countless Muslims.[30] For the question of succession must grapple with an *inability to be written* which devolves into internal fighting. Once again, fratricide shines a light on the formation of the political order, which is split between a repressed "civilized" part and a resistant "savage" part that precludes any possibility of writing, stoking the pleasure for murder. A relentlessly active blank space is put into action. And so two matters enter into conflict: the *forgotten word* of the Prophet, who has named his successor, and the *impossibility* of attesting to this pronouncement in writing. The two memories of the psyche and the political order are thus actively at play in the religious field.

In an inchoate, memory-less present, writing in blood (springing from war) takes the place of the dismissed textual testimony. What appears to be an absence of History and memory in the form of narratives is actually an overload of memory drowned out by a blank space and thus illegible and unattainable. *Forgetting is at play, but what traces there are find themselves accompanied by something else: the disappearance of archives, or their hijacking by another memory.* What empty spaces do we have at our disposal to find a little bit of freedom? Why and how is a History over-determined by blank space a form of enslavement that only pushes freedom further and further away?

9

Getting Out of the Colonial Pact

– Look what that bitch makes us go through!
– Who are you talking about? he asked Bouzid.
– Freedom.

Malek Haddad, 1989[1]

Don't invent new wounds, but new depths to your smile and joy: the world is there, drawn by your gestures like the starlight by your hand.

Nabile Farès, 1976[2]

Still today, "Algeria" – signifier, object, and place – is associated with devastation and upheaval, while the long-term effects of its colonial history are repeatedly disavowed. This is a consequence of a coloniality that strove to stop time, compress space, and erase memory for all its members. In France, this history structures discourse and, more notable still, its heavy silence. Colonialism and its traces look on the surface like an absence of memory, or, to use Daniel Mesguich's expression regarding the War of Liberation, like memory's "vast blankness" (see chapter 3 above). These unuttered murmurs concerning "Algeria" imply that there remains an unnamed, but fully active, inability to forget. And this is why it is still difficult after several generations to historicize this particular past and to create a narrative that respects its memory. A violence persists that is at once mute and deafening, erased by blank space. Does it have to?

The unrelenting difficulty of documenting the past threatens to give way to forgetting. As I've shown, coloniality creates a strange phenomenon whereby, as the novelist Salim Bachi has expressed it, "memory is born where there is an absence of traces."[3] At stake is a paradoxical frenzy of remembering which embeds History in the political order: the work of historians is prevented from setting off a debate that would impact the general public. Passion for "Algeria" continues to haunt the political order, in part through a "nationalism" which, devoid of any political agenda, only has its "unconditional love" for the fatherland. Any act, no matter how insignificant, that is viewed as deviating from the "national" cause is treated as treasonous and a return to colonial rule. *Hogra* persists in the imaginary as a catalyst for thought and social harmony.

After Liberation, the indefatigable re-enactment of coloniality within subjectivities and the political order

Coloniality invites civil war. On several occasions, the political apparatus in France that organized colonial society began to fracture over the question of Algeria. Let's recall that, for the "metropole," the so-called "Algerian war" was also a civil war, as the colony had become a part of the Republic. It had also caused a major rift within the French political order as civil war threatened to break out on its own soil. The feared confrontation did take place, but on Algerian soil in a war fought against its own people, first in the summer of 1962 and then again with the Internal War of the 1990s.

An agent of coloniality, fratricide thus entertains a dangerous relationship with the political order. This can be seen today in the Middle East, which, in the name of "democracy," is witnessing instead an undying empire. Capitalism ensures its own eternal preservation at the cost of civil war and tribal fighting. This also confirms the fact that, as a consequence of coloniality, fratricide lies dormant within democratic societies. The current logic of empire repeatedly afflicting the Middle East is closely followed and scrutinized by the subjects of former colonies for whom the ties between coloniality and the political order are far too apparent. Among the former colonial powers, meanwhile, these ties often go unnoticed, as they are blinded by their good intentions to bring "democracy" and by their belief, still after all these years, that they can civilize the "world" through democracy, at the same time ignoring the historical conditions that have given rise to their own political order.

It is thus easy to understand why, since the 2000s, the hawkish discourse of Western leaders and the frequent attacks on the Middle East have convinced the people of Algeria that the repeated evocation of "democracy" signals no more than the return of coloniality. Among other things, this has had the effect of reinforcing mistrust among the general public in response to any calls for democratization. Fewer demands are now made for political plurality and for a robust citizenship. The possibility of ending certain forms of servitude once and for all is met with fear among many in Algeria that this will only lead to a sort of *made-in-France*[4] "democracy." For coloniality – it bears repeating – found support by drawing on the Declaration of the Rights of Man and of the Citizen. This remains an unforgettable reality in Algeria, as this document has long served as a cover for other political goals: its claim of universality has been nothing more than a political instrument of oppression and justification for various forms of segregation. Many are, to say the least, very wary of universalism and the "democratic" pact as they have been put to use by the French political order.

The current situation in the Middle East thus reanimates Algeria's unforgettable colonial history. Every structural change of the political order bears the risk of devolving into a myriad of "worse-off" situations. In 1959, Frantz Fanon shared the remarks of a Swedish journalist who visited a camp of Algerian refugees. She spoke of "a boy of seven marked by deep wounds made by a steel wire with which he had been bound while French soldiers mistreated and killed his parents and sisters. A lieutenant had forcefully kept the boy's eyes open, so that he would see and remember this for a long time." Fanon then asks: "Does anyone think it is easy to make this child of seven forget both the murder of his family and his enormous vengeance? Is this orphaned child growing up in an apocalyptic atmosphere the sole message that French democracy will leave?"[5] How can one escape this unforgettably violent logic that has instilled fear and terror in people for generations to come?

On the one hand, in Algeria, coloniality has become a historical touchstone: it continues to occupy subjectivities as well as the political order, and is labeled without second thought as the cause of every ill. On the other hand, in France, coloniality is disappearing from History while still wielding power by creating blank spaces in political discourse and practices. An example of this is the way the political order in France has strived to unburden itself of its colonial history, preferring to cast it as a matter of concern only for former "colonized" subjects. In both places, coloniality thus continues its work of blanking out through erasure, disappearance, and dispossession.

In Algeria, the question of dispossession is at the center of governance and social relations. Colonialism is re-enacted by the political order, but it has especially marked subjectivities, regardless of an individual's social standing. Dispossessing the Other of a supposed potential he or she may possess remains a regular occurrence in interpersonal relations: it isn't only a question of seizing power with the aim of gaining a specific advantage, but also a matter of dispossession for the sake of dispossession. It is as though time has halted around the incessant act of stripping the Other of something he or she possesses. Therein lies the real invisible trace of colonialism, which spreads unchanged from generation to generation.

The political order thus translates the conflicts plaguing subjectivities into acts. Conversely, as citizens, subjects take part in constructing the apparatuses of power, as Frantz Fanon made clear as early as 1961: "A government and a party get the people they deserve. And in the more or less long term a people gets the government it deserves."[6] Fanon's warning brings to light the way citizens adapt to and reshape the decisions made by a political regime, even when they constantly denounce it and decry "treason against the nation." Equating those holding power in Algeria today with colonizers of days past falls short of understanding the underlying problem; rather, one must consider the way the subjectivities of citizens are themselves agents of subjugation, a subjugation they encourage as they dispossess their own people in a re-enactment of colonialism.

This is why the freedom won doesn't amount to an end of colonialism. Independence can re-establish a mode of colonial life that reorganizes the social bond. Freedom is a fundamental step toward experiencing an independent existence and citizenship. But this freedom can reverse course and become a refusal to separate from the traumatic moment of colonial rupture. Some obstacle prevents the emergence once and for all as a separate being who can be part of History without being subsumed by it. The non-separation with the colonial *spirit* makes History a current event.

The specter of colonialism returns to invade psyches and the political order. This is of singular importance, since, in both Algeria and France, the confusion of memory aids in the refusal of separation, which prevents the passion for "Algeria" from being treated. Erasure and the struggle to document the past give this passion its supreme power. Dispossessed of History and a sense of responsibility, the colonial order is thus perpetuated in the contemporary period. Does this mean that both the colonizers and the colonized are involved in this? How, then, can one let go of the colonial gesture of dispossession which functions as memory and History on both sides of the Mediterranean?

The work of the historian can never gain its necessary autonomy, as it is either stripped of political import in France, or appropriated by the political order in Algeria. As shown, dispossession in Algeria laid waste to core symbolic references, which were replaced in the minds of many by blank space. This mechanism that is ceaselessly at work is an act of memory, celebrated and shared by all the members of the colony, even if the crimes, destruction, and haunting of disappearances remain fundamentally unshareable. Both in France and in Algeria, this matter moves through time and space unchanged. Is this some strange "colonial pact" (Frantz Fanon), *invisibly* signed without anyone's consent or awareness? And what is one to make of the asymmetrical base of a supposed symmetrical pact? Indeed, if this pact was established between colonizers and colonized, neither of them entered into it on the same level or at the same time. The colonial rupture was unprecedented in its destruction and impact on future generations. But now in hindsight we can ask the following question: what did the colonized group agree to by re-enacting colonialism after gaining its freedom?

An overly radical reading of Frantz Fanon has downplayed his invaluable contributions to psychiatry. And yet it is from his clinical work with both "*indigènes*" and "Europeans" suffering from psychological disorders (from 1953 to 1956) that he began to consider the positions of colonizer and colonized as agents of a system and not as simple executors or victims.[7] Thus, when he writes in 1961 that "there is not one colonized subject who at least once a day does not dream of taking the place of the colonist,"[8] he offers a new model of understanding. For, lacking the ability to remember, which can situate the matter as the product of a shock felt in the past, envy is the corollary of dispossession. Re-enactment in fact emerges in the absence of textual memory, which would otherwise halt it. With core symbolic references destroyed, inscription, having no matter of support, disappears in turn.

The space of core references (and History) was wiped out and lineages were erased. All this to make an entire people disappear. This "genealogical wound"[9] constitutes an attack on the symbolic order, which holds body, language, and psyche together. Colonialism's destruction of ancestral space plunged each "*indigène*" into various forms of melancholy and abandonment: "I pushed open the door of the space," Farès writes, "and something broke in me. Like a plank. Or a pleasure. Inanimate. I pushed open the door of the space and I was able to enter into the time of my existence, as the inside had just split open."[10]

Trauma as shelter and alibi

If core references are undermined, interiority itself becomes "fissured" and thus susceptible to becoming occupied and dominated. After this, time stands still, immobilizing at the initial moment of rupture, which, lacking any other reference points, becomes unforgettable. *Rupture, dispossession, and disappearance end up over time becoming history*. Colonization begins operating like an origin myth. In spite of Liberation, it becomes the ideal foundation for the subject and the political order. More precisely, this concerns a political order that becomes decidedly apolitical. In other words, the Algerian subject is dispossessed of political agency. The revolutionary period brilliantly distills a political potential that, as soon as the Revolution is over, turns into its opposite. It abandons its subversive dimension and goes "through the dark door of refusal" (Jean El Mouhoub Amrouche).

Colonization in Algeria is both a myth and a distraction for any real work of subjectivication, which offers the only path toward subjecthood. It works by inserting itself into any reading and interpretation of this episode of History. Is there any better way of agreeing to one's own subjugation and encouraging the occupation of one's interiority – in this case while disdaining freedom, that "bitch," as Malek Haddad called it?

This line of thinking reinforces on a massive scale colonialism's absolute power. The Other of colonialism – and then later the post-Independence state – is granted infinite and unassailable dominance. The authority given the Other helps to distract from the hollow core of every totalitarian system (including that of colonialism). The subject believes in the Other's absolute power; it is a matter of faith, functioning almost like a religion. And so, if an all-powerful Other exists, "I," as a subject, risk going beyond my limits. Colonization, as historical fact and origin, places a low ceiling over the subject and the political order. It blanks out memory, "compresses" time and space. Move on, there's nothing to see here! It is easier to cling to the destruction caused by the Other than by that caused by oneself. Colonial occupation shields and surreptitiously attacks at the same time. It is difficult to step out from behind its protection. The dispossessed subject is in a state of fascination, at once completely hollowed out and fully identifying with that which dispossesses it.

The "colonial pact" makes itself felt on the level of the real when the "colonized" cling to what they thought they had finally freed themselves from. They continue to keep something of the occupier alive (new appearance notwithstanding) in order to fight against the latter's disappearance, which remains a threat to ward off. This allows us to understand the longstanding struggle in Algeria for the country to view itself as a collective whole. Occupation provokes rupture and unrest. The death drive undermines the social bond, which succumbs to melancholy with endless, though hollow, grievances. Civil society struggles to take shape despite the immense effort of individuals. It remains impossible to move past the destruction of the land and of social ties. And if a collective whole forms among individuals, there is a great risk that another individual will try to destroy it in the name of the "colonial pact."

Examining neuroses caused by war, Sándor Ferenczi coined the term "unconscious traumatophilia."[11] The traumatized subject makes its shock, to which it is viscerally attached, the center of its existence. It begins to grow fond of its trauma like its "own" abode which bears the traces and marks of this internal conflict. Failing to see the Other as welcoming and well meaning, this attachment to destruction occasions a return to a primitive narcissism. Alterity remains threatening for the traumatized subject, as its unforgettable wound came from being exposed to an outside danger. The subject thus takes refuge within its own people. It fully inhabits its wound, which serves as an archive for the unforgettable.

Similarly, the colonized part of the subject is precisely the one that remains attached to a form of inertia and a particular pleasure derived from the perceived protection of dispossession. The subject fascinated by its dispossession takes refuge in origins and in a narcissism from which it struggles to escape. It likes the Other, who is behind its destruction, and doesn't cease to wonder over the Other's awe-inspiring power, even while the subject continues to voice grievances about it. The subject ends up unwittingly agreeing to practice dispossession within the collective. Allowing History to remain at a standstill, the "colonial pact" continues to rely on dispossession, erasure, and disappearance, confirming Fanon's idea that "colonialism ... has incrusted itself with the prospect of enduring forever."[12] Between colonizers and the colonized there is a paradoxical relation to time. Colonizers enter into a time that lasts for a hoped-for eternity – as the writer Louis Bertrand made clear, recalling how the colonizers, upon discovering Roman ruins, were struck by the supposed Latin pre-history of their own presence. The colonized, for their part, are stuck in a suspended time that is frozen on the exact moment of their trauma. The subject is besieged from within by colonial dispossession to such an extent that it can't tell the "invisible" occupier from its own skin. It thinks, embracing the myth of the colonizer, it can reach eternity this way. Indeed, what is more eternal than shadows?

The "colonial pact," as shown in the previous chapter, was built upon a myth that occasions both birth and destruction (of the subject and the nation). But it is also a myth that strips the subject of all responsibility for its actions, its words, and its history. This is because responsibility is always the matter of an elusive Other, the other party who agreed to the pact, whether that be the colonial power or the so-called "postcolonial" state. In either case, the subject is exonerated from participating in History or politics. It rages against the fact that it is still dispossessed while it unknowingly performs its own dispossession. It becomes its own oppressor and inner enemy in honor of a traceless, text-less pact it no longer remembers. It even divests itself of any political agency to give full power to the state, whose practice of disappearance and totalitarianism it decries. And all this in a context where the political power has made a calculated move to embrace the specter of colonialism. It has done this by both embodying its haunting presence[13] and by extending colonial impunity through the so-called law of "national reconciliation." Let's recall that, in the context of the Évian Accords, the French military and political leaders were granted amnesty for the crime, torture, and atrocities they carried out, allowing amnesia to erase the political order's relation to its own history. We have already seen how the political power in Algeria had also orchestrated its own impunity during the years of the Internal War.

Since the beginning of conquest, erasure continued to be a weapon used under occupation. It was first a question of erasing the practice of disappearance (from memory) of almost a third of the population, then of erasing genealogies by forcing the question "Who is who?," and finally of erasing the sense of responsibility. But another erasure is also at play, as can be seen in a statement made by Victor Hugo in 1879, as recounted by the historian Gilles Manceron: "Africa has no history; it is covered by a vast and obscure legend," he said, while pushing for a "peaceful colonization" and seeking to justify the invasion of Algeria: "People! Seize this land. Take it from whom? From no one. ... Take God's land ... for the sake of industry; not in the spirit of conquest but in the spirit of fraternity."[14] As we have seen, in post-Independence Algeria, the role played by fratricide in the history of nationalism was also erased so as to deflect any questions of responsibility, pointing the finger at the colonial Other in order to distract from its own responsibility. The embrace of Arab-Islamic colonization also aimed to erase French colonization by performing an exact substitution, which offered the advantage of maintaining an already established occupation.

In Algeria, the practice of disappearance was used as a weapon during the Internal War (1992–2000), reinstating an "atmospheric" terror (Fanon) while few connected this to the long history of this practice under colonization. Disappearance haunts memories by blanking them out and causing the subject to doubt its own existence: "Do I really exist? Am I alive or dead?" The haunting of disappearance is widely expressed as a feeling of being crushed under the weight of invisible forces. It isn't uncommon to hear in this respect: "I am the living dead." The subject is thus forced to bear the ghost of the state and its supposed immortality. Conversely, the state embodies the subject's ghostlike manner, allowing the latter to see itself in the former. The melancholy this provokes slowly eats away at the social body. A recent crisis serves as a telling reminder of how the living remain plagued by the disappeared and haunted by their own disappearance.

The brutalization of the living: the disappearance of children

Since the beginning of the 2000s, children from eleven months to ten years old have regularly gone missing in Algeria. The number of children involved and the frequency of the occurrence are alarming. This isn't an isolated fact, but a serious symptom of the death drive affecting the public at large. The pain widely felt is evident almost everywhere but hardly recognized. The absence of serious studies of this all-too-recent phenomenon gives no option but to rely on the information put forth by the Algerian press – which should be regarded with caution. In most cases, according to journalists, children are met with the same fate: a girl or boy is abducted by one or more adults, most often someone close to, or part of, the family – disappearance is re-enacted within one's own family and community – sometimes by an unknown person or people. Most of the time, the child is never seen again, provoking in psyches a horror tied to the absence of traces.

According to some journalists, an estimated 848 children were abducted between 2000 and 2006, 86 of whom were found dead, cut into pieces or showing clear signs of sexual abuse; the others seem to have disappeared without a trace. In 2013, 200 children were reported abducted, and in 2014, 195. In September 2015, a journalist reported a total of "a thousand children abducted over the past ten years and 50,000 children abused per year, with 10,000 of those sexually abused."[15] "Each year," a journalist for the French-language daily *El Watan* wrote a month later, "hundreds of children disappear, they are abducted then killed. In 2015 alone, more than 250 cases of minors were reported abducted across the country. And sadly today there is a steady stream of missing people alerts and other reports of children who left their homes and never returned."[16] In August 2016, another journalist claimed there were 5,000 children abused in 2015 alone, not including those who suffered sexual abuse.[17] There were also an estimated 250 cases of abducted children in 2015 alone throughout all of Algeria.

The numbers reported by the press are thus rough and contradictory estimates. There are no unanimously accepted and precise data on this matter. Once again, the absence of documentation makes itself felt, making it hard to even find the facts, much less interpret them. All one can do is infer from the journalists' approximate accounts that the horror corresponds to facts that cannot be verified. It therefore remains difficult to measure their real impact.

The state of recovered bodies – decapitated, savagely hacked to pieces after suffering sexual abuse, according to the press – indeed provokes horror. The living are thrust into a state of shock at this display of butchery. All the more so given the fact that these are children, which is to say, vulnerable, fragile beings at the mercy of adults, who, regardless of their relation to the child, are supposed to look after them. A strange reversal takes place between child and adult. The adult goes from being the one who ensures the child's well-being to becoming a potential threat to it.

As for the children, their place is entirely upturned in a Machiavellian manner: they lose their status as *maléïka*, untouchable angels, and become objects of consumption destined to be cut up, raped, missing. In the past, however, children had a well-defined role in various traditions and customs found throughout Algeria. They were the link between the divine and the ancestors. On the one hand, by being deemed *maléïka*, they were to be protected by humanity by divine order and acquired in this way a certain power that made them "untouchable." On the other hand, it was common to name children according to their gender, *Baba* (father) or *Yemma* (mother). This placed them right away in the place of ancestors to be loved or feared. In the popular beliefs that people subjectivities, the child, this "father of Man" (Freud), was a sacred figure. So what happened to make the child assume for many over time the position of a worthless object, condemned to the worse outbursts of hate, violence, and destruction?

Disappearance, sexual abuse, dismemberment of bodies, and the way some bodies have been discovered (in trash bags, at the bottom of wells) attest to the treatment of the body as a worthless object, worse even than a corpse no longer worthy of burial. By destroying what the child represents – promise of life and future hope, link between different generations and time periods – a shift has occurred in the very relation to death, and, by extension, to life. This is a question not simply of the brutal killing of children, but of a much more consequential brutalization of the living, which leaves open wounds and provokes panic on the level of the imaginary. Regardless of their linguistic and cultural backgrounds, human beings all possess belief systems and rituals for confronting the inexplicability of death. But how can one explain this cruelty unleashed on children, where History "becomes flesh"? How can the living be concerned with anything apart from disappearance and mutilation?

The "bone seekers": from children to fathers

In many ways, history is defined today by the headline "missing child," which forces the living to become "bone seekers," to borrow the title of Tahar Djaout's wonderful novel. In this 1984 work, Djaout recounts how, in the immediate aftermath of Independence, an entire population of a Kabylie village, pickaxe in hand, goes in search of "bones," the remains of the missing from the War of Liberation: "My brother's bones are waiting for us like a treasure, buried among other heroic cadavers over which orations and praises breed like worms attracted to a rotting carcass."[18]

Djaout brings together the two levels of this search while calling attention to its unlikely humor: the urgent and profoundly human need to provide a burial for one's own, and the need to show "corpses" as a sort of proof. The proof is needed to attain the status of the family of a martyr (*chahid*) and to benefit from the financial, material, and/or symbolic advantages that come with this: "But people were attached to their dead as if to an irrefutable proof to be displayed one day before the betrayal of time and man."[19] The martyr is an (immortal) missing person who redefines the lives of the living. In this novel, Djaout reveals the terrible power the dead hold over the living, especially when those gone missing were never granted funeral rites. The missing become wandering ghosts, endowed with mad, dreadful powers.

The village men go seeking the scattered remains of corpses to give them the semblance of unity and a dignified burial. In order to do this, they have to keep digging over and over, with the risk of finding bone after bone that doesn't belong to any identifiable individual. But these bones must be brought together, reassembled in order to recognize them with a dignified burial, which is to say, one with a patronym. The reader takes part in this horrifying search for remains, a quest aimed at turning the missing into the dead. It is a strange and shocking fact that, even today, there are families still "seeking the bones" of their children.

Seeking bones, unearthing them, fighting with and against disappearance time and time again, striving to give the missing a dignified death by giving them a name, a body, a story and finally a memory – does this still define our present moment? Are we still at war, and against whom? Are we in the midst of an unidentified war unfolding in a familiar setting among families, or is this simply the on-going ramifications of an ancient disaster? And finally, could the disappearance of children be seen simply as the continuation of state terror that characterized the Internal War and the practice of disappearance under colonialism? Isn't this what the law of "national reconciliation" seeks to erase from memory to uphold colonial impunity?

Worse than the disappearance of fathers, that of children knocks on the doors of the living. With these young beings named *Yemma* and *Baba*, we might say that the "ancestors" are coming to demand their due, disfigured by an intrafamilial savagery. The colonial use of disappearance as a weapon pays no heed to time or generational change: it is ceaselessly re-enacted within one's own community. From children to fathers and fathers to children, the living are rocked by an uninterrupted and silent history of disappearance. The missing loom heavily, threatening at all times to carry the living away into their world. The best response, then, is to make the missing children, these disfigured ancestors from generations past, part of an atemporal and unforgettable History.

Disappearance, destruction, and other forms of torture suffered by children demand as much. This phenomenon takes three to four generations to unfold. By reinstating these children into a genealogy and historical lineage, the re-enactment of History is exposed. The disappearance of children – and the cruelty they have had to endure – began in the 2000s.[20] In the 1990s, as shown, thousands of men, young and old, were taken away by police forces (see chapter 7 above). Following a generational pattern, they were potentially the fathers of the children who were currently missing and the sons of fathers who went missing during the War of Liberation. Still today, the living are striving to obtain information regarding the bodies of their missing family members. But they are left to live in dreadful solitude, without any judicial, social, or administrative support. They wander in search of tombs they'll never find.

Large numbers of "bone seekers" are still milling about, going to great lengths to ignore their own pain as they are driven by the imperious need to find out what happened to their loved ones and to recover their bodies. They have been crying out in despair for generations that death and disappearance are not the same thing, and especially that death cannot be decreed in place of another order of truth. *From fathers to children and children to fathers, it is the same story of disappearance. In other words, no matter what level one looks at (grandfather/father/child), one confronts the enigmatic abyss of disappearance and those who live with it, the bone seekers.*

Disappearance under colonialism, as we've seen, was a large-scale massacre whose case was dismissed. As a consequence, existence was dispossessed of life. The same situation reoccurred during the Internal War. And it is happening today under a different guise, which is just as worrying as it is more diffuse and the perpetrators remain unidentified. Is the repetition of History not subject to time?

Lacking burials, stories, and collective recognition, the missing fully occupy the site of memory. As Nabile Farès made clear in 1974, the current re-enactment of disappearance implies that "we have been bitten at our most tender places, and one day this will have to be fixed, the fish will need to go back to the sea it came from, and salt will have to act on the wound as the hook is tossed in a hole far away from us."[21] From generation to generation, what is kept quiet, silenced, or violently repressed returns in the very presence of the body and exists on the level of the real. Ignoring it or pushing back against it only makes this return all the more ferocious and therefore more savage. The living who are fully committed to finding their missing are themselves thrust into death or its other expressions: violence, inertia, despondency, suffocation.

Disappearance affects the living and the dead. It has little to do with absence, but instead concerns a ubiquitous specter. The power the missing hold over the living is immense, unlimited, and timeless. This seemingly eternal phenomenon was described by Tahar Djaout, who was also a victim of a faceless, but all-too-common, crime: he was murdered in June 1993 amid the confusion of "Who is killing whom?" Still today, his murder remains a mystery: was it carried out by the state or the Islamists? It remains unknown and his writing is all that is left. His novel *The Bone Seekers* can thus be read as a testimony of the living's plunge into melancholy as they seek their missing and the missing's unburied, dismembered bodies. Alas, the present rehearsal of the past with the disappearance of children has proved him right. Readers of his book steadily become word seekers, overwhelmed by the fractured bodies that can't be found. By the time they finish it, both readers and characters have been through a grueling test. What becomes of them? How can one remain living after a horrendous, terror-stricken search that has perhaps gone on for at least three, maybe four, generations?

> Traveling such distances and crossing so many villages reveal strange and difficult things about your fellowmen and about yourself. Without a word Rabah Ouali and I lie down under the olive tree as new luminous particles explode in the sky. Even the very natural joy of coming home after a long absence feels strange to us. *How many dead will actually come home to the village tomorrow? I'm sure that the deadest one of us is not my brother's skeleton, rattling inside the bag with unfeigned pleasure. Persistent in his efforts and his braying, the donkey is perhaps the only living being our small convoy is bringing back.*[22]

Conclusion: Ending the Colonial Curse: Lessons from Fanon

In order to grasp the gloom that plagues both individual subjects and the larger public in Algeria today, in order to understand and analyze the despair of subjectivities as expressed by a number of my patients, a deeper look at history and politics was needed. For the practice of psychoanalysis is also a mode of social interaction and, as such, it sheds light on the damage felt by subjectivities concerning their relations to themselves and those with others. The destruction of this relation over the course of thirteen decades of colonial domination, as we've seen, has plunged a number of post-Independence Algerian citizens into a state of melancholy that continues unabated today.

The social bond has been damaged and its ills have eaten away at the promise of social harmony. These ills have penetrated people's memory to the point of convincing them that there is nothing else but this to remember. And so memory and erasure co-exist, along with deep-seated grievances and denial (not to mention disavowal) and an extreme moralization of religion that can hardly be contained. The LRP (language, religion, and politics) apparatus is defined by these paradoxes. Its sought-after purity masks the untreated ills provoked by colonialism. And with colonialism having erased so much, there is no text prior to colonialism to turn to. Blank space fills the space left empty by erasure, with nothing there to document. However, unable to rely on a text pre-dating colonial destruction and offense, we can still ask what the formerly colonized are suffering from and the seemingly incomprehensible reasons they insist on remaining in this state: why – and how – is pleasure sought in prolonging the occupation of one's interiority by the specter of colonialism?

The "colonial pact": erasure of memory, disappearance of bodies, dispossession of existence

Identification with the colonizer, carefully detailed by Albert Memmi and a few others after him, cannot adequately account for the way subjects are penetrated by the spirit of colonialism, and it especially falls short in terms of understanding the feeling of internal disappearance. This is all the more the case in that disappearance, a colonial phenomenon, continues to threaten the living. Experienced historically, disappearance has been considered in the past decade in Algeria as something that can strike at any moment: fear of disappearing under colonialism, fear of disappearing under the post-Independence totalitarian regime, and then again during the Internal War, and still today with the economic crisis and the possible resurgence of political Islam or its offshoots. In Algeria, disappearance was experienced historically on the level of the real, but this past matter is constantly felt as looming. The mental space of the subject succumbs to occupation because something of itself has disappeared, and it has little left to lose. But while it is divested of its self, yearns to accumulate and fill itself up with "possessions." The occupation of its mental space, however, makes it believe it is full, satiated, and therefore lacking nothing. What's more, internal disappearance and its corollary, occupation, both inscribed by history, correspond to a mode of governance in Algeria: the execution of disappearances, on the one hand, and, on the other, the reign of specters. From one to the other, the blank space of the "colonial night" (Ferhat Abbas) continues to spread today.

Since Independence, as we've seen, the specter of disappearance has been exploited "to track" the living. Despite the re-enactment of traumatic events, there has been no attempt to engage the general public in the formation of a collective account of History. Subjects remain overwhelmed by the shock of disappearance. The public at large is prevented from dealing with these issues. In the absence of justice, the law of murder threatens to continue to wreak havoc. Over the course of several generations, subjects have been dealing with their dead and missing by becoming tombs themselves. From this results a real inertia in the social order, as the living, haunted by their potential disappearance, remain immobilized. Patients I have treated who lost someone during the Internal War struggle with an impossible grief. These men and women become the invisible burial vaults of their murdered loved ones. For these subjects, the dead are not dead. They wander about in their disappeared state, urging the living to join them in their spectral world. These subjects are plagued by a terrible feeling of being caught between a grave and a ghost in order to honor the undead dead. Once again, the political order strives to "nationalize" memory in order to discourage any independent initiatives to record traces. Thus, the LRP creates a confusion of memory and subjects are left to cope alone with their impossible grief. Let's recall how the law of "national reconciliation" is in reality a means of disavowing crimes and perpetuating the widespread practice of disappearance.

Everything is in place to uphold the "colonial pact," as people continue to go missing and criminals go unpunished. Subjects are besieged from within by the thought of disappearance. This is where the real "colonial pact" finds expression, forcing the living to become mere shadows of themselves. The living are caught under a troubling spell: how then can they move on from mourning the missing without a collective burial? How can they give names and bodies to the missing when, in the name of the law – a state of exception under colonialism or a state of emergency in post-Independence Algeria – efforts are made to make the missing themselves disappear? Does this erasure of disappearance's existence on the level of the real not represent the worst side of all the violence?

Erasure of memory, disappearance of bodies, dispossession of existence – such are the signifiers of the real "colonial pact" that aims to maintain subjects "under its spell" (Fanon) in an apolitical state. Possession is a signifier of madness. Many traditions have long understood that madness comes from being possessed by a spectral world that eats away at the subject from within. In both Arabic and Amazigh, the mad are indeed described as being inhabited, under occupation (*meskoun*). Their psyche no longer belongs to them, it has been taken over by demons (*djinns*). At the height of the War of Liberation, Frantz Fanon, whose psychiatric practice considered both the individual and the public at large, began studying the rites and rituals of exorcism among the Marabouts.[1] To be sure, he wished to study these rites and gain a better understanding of the traditional practices of "*indigènes*" in order to better treat them. But there is another level to this interest, too, one that springs from his intuitive understanding of the impact that colonialism had on all subjects who had long suffered under its law.

The mystical quality of the colonized

As early as 1960, Fanon clearly perceived that Independence didn't imply a break from servitude. This is why, for him, the "nation" needed to redefine itself in order to move away from the signifiers of servitude that had led to the insurrection. Anything less and occupation and possession would be used by a system of governance to control the masses under the idea of doing what is "best for the nation." And this logic of political power did in fact come to reign in post-independence Algeria by casting a spell over subjects in an effort to maintain the "colonial pact." These spellbound subjects wander about like sleepwalkers, acting as ghosts or the living dead in search of a body. In this way, the "nation," Fanon affirms, becomes an instrument to cast a spell on people, enshrouding them in confusion and stirring up their emotions.[2] The war rages on in a context where everyone becomes an unidentifiable enemy. This speaks to colonialism's grip on the bodies of subjects, earning their consent by eliminating any hint of insurrection and driving them into a form of passivity. Is the possession of the colonized subject and its obsession with a spectral world mistaken for a mystical state?

In a 1957 article on "The Phenomenon of Agitation in the Psychiatric Milieu," Fanon argues that the "breakdown of the [subject's] real world" allows it "to hallucinate."[3] In other words, hallucination can have a temporarily beneficial effect on the subject suffering from trauma: in the absence of interiority, the subject moves into a world peopled by ghosts until it is completely taken over by these guests. Fanon thus believes that understanding trance and possession is essential for understanding colonialism. For dance and rituals for warding off a spell make the body move and abandon the rigidity of the colonized body, which is to say, the occupied body. During these ceremonies, the group welcomes the subject's occupiers, its *djinns*, in order to separate the possessed from those who possess it. The subject's liberation is thus just as much a collective affair as it is an individual one: no liberation is possible for the public at large without liberating the subject, and, conversely, the subject's liberation requires that of the larger public. And so, for a practitioner of psychic care such as Fanon, collective liberation is also seen to be a decision made by each individual subject.

Following the lessons of Freud and Fanon, the subject is formed by being "left to himself" (Freud)[4] or "let go of" (Fanon). This means that the subject's position is "supported" by a radical solitude. If freedom can only come from the Other of the larger public, it is also incumbent upon each subject to pull itself away from it in order to liberate the public at large. But some modes of collective organization squash the very possibility of liberation within the private sphere. This is where many subjects treated by psychoanalysis struggle and put up resistance: they complain that their mental space has been besieged and yet they resist an inner revolution. And since everything in social life only reinforces the law of "mental" occupation, they prefer to obey it religiously. Here again, the haunting of disappearance makes itself felt: freedom as a liberation of mental space is imagined as a form of disappearance. The very same disappearance that has shaped history.

The body continues to be immobilized, thought functions in a low-key mode, and desire is frozen. This rigidifying process is a symptom of a terrible fear provoked by disappearance. In the end, the haunting of the mind makes existence feel painful. This can be heard in an expression that was frequently used in Algeria during a "housing crisis" when affordable housing was scarce: *Andi el rachi fi rassi* ("My mind is filled with people"). It is worth noting in passing that religion perfectly performs this function of occupying interiorities, dictating the subject's behavior in order to spare it the solitude of its actions.

Following the thinking of Yamina Mechakra in *La Grotte éclaté* (see chapter 3 above) and that of Frantz Fanon in *The Wretched of the Earth*, it seems the colonized part of the subject seeks its own disenchantment. To do this, the larger public must be both the healer and the stage where different performances of the besieged subject can be played out. Fanon notes that, during the war years, traditional practices of dispossession greatly diminished because violence was released in other ways related to war. There is nothing exotic about this observation of Fanon's, but there is a mystical quality to it. Here Fanon is advancing two arguments: the first, that colonialism is defined by its occupation of spaces, notably that of "mental" space; the second, that decolonization can only take place via the subject's self-liberation, which depends on the approval and participation of the larger public. It is thus a matter of creating and allowing for catharsis-provoking scenarios. This is what Kateb Yacine and Abdelkader Alloula had set out to do. Alloula was killed during the Internal War.

Staging performances and other artistic productions allows for the invention of liberation. Proposing a performance of exorcism as a form of liberation may seem odd. And yet what is needed is a collective effort to free the spirit of the disappeared, through our words, through justice, and through other forms of artistic production. There is still a long way to go.

"Any study of the colonial world," Fanon writes,

> therefore must include an understanding of the phenomena of dance and possession. The colonized's way of relaxing is precisely this muscular orgy during which the most brutal aggressiveness and impulsive violence are channeled, transformed, and spirited away. The dance circle is a permissive circle. It protects and empowers. At a fixed time and a fixed date men and women assemble in a given place, and under the solemn gaze of the tribe launch themselves into a seemingly disarticulated, but in fact extremely ritualized, pantomime where the exorcism, liberation, and expression of a community are grandiosely and spontaneously played out through shaking of the head, and back and forward thrusts of the body. Everything is permitted in the dance circle. ... Everything is permitted, for in fact the sole purpose of gathering is to let the supercharged libido and the stifled aggressiveness spew out volcanically. Symbolic killings, figurative cavalcades, and imagined multiple murders, everything has to come out. The ill humors seep out, tumultuous as lava flows. One step further and we find ourselves in deep possession. In actual fact, there are organized seances of possession and dispossession: vampirism, possession by djinns, by zombies, and by Legba, the illustrious god of voodoo.[5]

These rituals designed to break colonialism's spell helped balance the libido of the subject and the community. They gave rise to a liberation that was otherwise forbidden. How can one invent a practice of exorcism that doesn't fall into mysticism?

For a future liberation

Fanon's work has many facets, each of which opens different lines of inquiry and analysis. His work straddles many different disciplines: history, sociology, anthropology, political philosophy, and psychiatry. His writing eludes classification and cannot be confined to a single intellectual, regional, or linguistic tradition. His thinking casts itself as outsider, in a perpetual state of self-exile. His thought emerges from a vanishing point where fields of knowledge and discourse continue to unfold endlessly. Fanon's writing seeks to bring about a decolonization of knowledge and practices.[6] For Fanon, freedom is not conceptual, but an act of emancipation through writing. This reading of Fanon should prevent his writing from being taken hostage by a militant position.

And yet his thinking is far too often read in this reductionist manner, probably in order to blunt its edge. But this edge is what gives his thinking far greater reach than a militant reading suggests. Fanon the militant was only putting into practice his thinking as a psychic care practitioner. From militant to psychiatrist, there is no significant shift, but rather a continuity from thought to action. Fanon drew on his clinical practice at the Blida-Joinville Hospital to develop a theory of the subject that goes beyond the colonial situation. To my knowledge, neither in France nor in Algeria is his work taught to aspiring healthcare providers specializing in psychic, psychiatric, or psychological work, nor is it used to train psychoanalysts. Yet Fanon's thinking offers an invaluable clinical reflection on freedom on both a personal and a collective scale. He clearly shows how the subjugation of the colonized subject gives rise to an occupied mental space that has a major impact on its personality: this "occupation" masks the subject's fundamental alienation, the one inherent in all subjects regardless of their historical, linguistic, or political origins. The colonized subject is thus deprived of experiencing its own alienation by being subjugated by the Other of colonialism. Dispossessed, it becomes possessed. Which makes it easy to understand why the subject is rarely responsible for its actions.

To be a subject, for Fanon, is to be free. This is why he refused to continue his practice in a context of colonial oppression. He couldn't help liberate subjects from their troubles in a system that created "depersonalization" and "decerebralization."[7] Fanon speaks of these troubles as "pathologies of freedom": "In any phenomenology in which the major alterations of consciousness are left aside, mental illness is presented as a veritable pathology of freedom. Illness situates the patient in a world in which his or her freedom, will and desires are constantly broken by obsessions, inhibitions, countermands, anxieties."[8] Freedom implies and determines a subject's function. In this perspective, Fanon's fight against colonization is a matter of translating the clinical theory of the subject he developed as a psychiatrist into the political arena.

It is worth pointing out that two opposing conceptions of madness and freedom are implicitly at play here: that of Fanon (and found in Henri Ey) and that of Lacan (developed in opposition to Ey). For Fanon, madness is an obstacle to freedom. For Lacan, it is the very space of freedom. The fact remains that, for both, the subject is indissolubly bound up with politics. In his "Presentation on Psychical Causality"(1949), Lacan writes:

> Thus rather than resulting from a contingent fact – the frailties of his organism – madness is the permanent virtuality of a gap opened up in his essence. And far from being an "insult" to freedom, madness is freedom's most faithful companion, following its every move like a shadow. Not only can man's being not be understood without madness, but it would not be man's being if it did not bear madness within itself as the limit of his freedom. ... A weak organism, a deranged imagination, and conflicts beyond one's capacities do not suffice to cause madness.[9]

And, he adds, "Not just anyone can go mad." But are these two conceptions of freedom and madness really at odds?

It seems to me that in Fanon's thinking there is a dilemma tied to the colonial context in which he develops his theory. He envisions freedom as an act of emancipation from colonial perversion and, for this reason, he becomes politically engaged. And at the same time, becoming a politically engaged theoretician of the Revolution, his call for freedom entails in part a leap of faith, which he himself is skeptical of as he warns revolutionaries about the risks of a reversal of freedom after Independence. Fanon's writing thus gives expression to two meanings of freedom: the first affirms freedom here and now, existing on the level of the real as an act of emancipation and a refusal of subjugation; in the second meaning, freedom remains a goal on the horizon and, as such, entails faith and patience as it is sustained by the imaginary.

If Algerian Independence occasioned a return to subjugation, it is also because freedom isn't enough to *make free*. To be sure, freedom depends in part on its standing within the imaginary, but it can only lead to a genuine emancipation, to an act accomplished by each citizen, if it is endlessly fought for as a potentiality, always on the horizon. Freedom was taken for granted in the depoliticized period following the Revolution, and, consequently, no one hoped for it or fought for it.

Developing with others instruments of culture and knowledge, reinventing one's alienations in the hope of breaking free from them: this is the new urgency in today's Algeria. Otherwise, the endless cycle of economic, social, and political crises threatens to drag citizens back into the abyss that has been there all along: the abyss of melancholy. Now is the time to find the means of turning trauma into a fruitful source for thought and political action. And to create spaces for memory for future generations that break the cycle of re-enactment.

For Fanon, freedom isn't determined by historical, biological, social, or economic factors. It is determined by someone who steps up and proclaims, "in a broken voice":

> There is no white world; there is no white ethic – any more than there is a white intelligence. There are from one end of the world to the other men who are searching. I am not a prisoner of History. I must not look for the meaning of my destiny in that direction. I must constantly remind myself that the real *leap* consists of introducing invention into life. In the world I am heading for, I am endlessly creating myself. I show solidarity with humanity provided I can go one step further.[10]

Freedom cannot be conceptualized, even less controlled. It is experienced in action. It remains fragile, an excess of the real, threatened to be captured by the reign of faith. Freedom becomes possible only when the subject accepts its own alienations. And freedom remains in a constant state of becoming. These two levels of freedom only give rise to liberation if they are forever held in tension. In Lacanian terms, we might say that freedom – the sort that Fanon sees as behind his own existence – is the *real part of the living subject*. It is there, out of reach, yet present when speech becomes subversive. The Fanonian contradiction between freedom and madness contains its own resolution if one accepts, and assumes, his own words on the matter: "There should be no attempt to fixate man, since it is his destiny to be unleashed."[11]

Born of a monarchical system, coloniality works first and foremost by seizing bodies – physical, symbolic (language), and imaginary (legends and myths). And from this step, the spirit of colonialism violently penetrates psyches, which means that, in spite of the generations that have come and gone, Independence still hasn't broken free from colonial power. As a result, in Algeria, freedom has become an act of refusal that each subject re-enacts at its own expense.

But let's not forget that this unthinkable separation *concerns both countries*. Algeria first and foremost, of course, where the occupation of "mental space" was designed by colonization to last, such that the harmful effects of the "colonial pact" continue to be felt. But also in France, where most are in denial about this occupation, even though the country's history in Algeria is constantly implicated in political discourse and by individuals. Coloniality aimed to embody the Other, including by making it its waste (the "*indigène*"). More than half a century after the "end of colonies," the descendants of the colonizers and the colonized still struggle to break free from the spirit of colonialism and to reclaim their ability to think independently – which was confiscated by the political order – and apply it to history. The time has truly come for a plurality of memory to emerge peacefully from History, one in which each subject could recognize itself, feel welcomed and free to exist. At last, humanity could deliver its "broken" song, and History could finally become subject to public debate.

Notes

Foreword

1 In this connection, I quote from my book *El trabajo del testigo. Testimonio y experiencia traumática* (2016): "Jean-François Lyotard wondered whether it was the historian's task to attend not only to the damage of history, but also to the destruction of its documents. ... Here there is a painful analogy with the disappeared in Argentina (where to disappear a person was also 'to kill death,' as Gilou García Reinoso wrote in 1986), who leave their traces in testimony as 'disappeared' and not only as 'dead,' perhaps without the law's being able to ask after this distinction" (p. 88).
2 *Translator's Note*: I have used the gender-neutral "it" to refer to "the subject" throughout this foreword, both to convey the general nature of this category and to preserve the genderlessness of the possessive pronoun in Spanish.
3 The work of Silvia Bleichmar (an Argentine psychoanalyst who died prematurely in 2007) has been translated into French and Portuguese but is not well known in the English-speaking world. Bleichmar's prolific work has been foundational in the Southern Cone, both because of its approach to the processes involved in the subject's constitution and because of its construction of a metapsychology that sheds light on the interconnections between the political and the subjective, without losing sight of an ethical dimension that is constitutive for the subject.
4 Many psychoanalysts in Argentina, myself included, have engaged with the work of the French psychoanalyst Piera Aulagnier, not

only in an effort to give an account of the constitutive matrix of the infant, but also as part of the work of thinking through the subjective effects of political phenomena. This means questioning our own practice and our ties to psychoanalytic institutions.

Introduction: The Difficulty of Acknowledging Colonial Trauma

1 Frantz Fanon, "Psychiatric Writings," in *Alienation and Freedom*, eds. Jean Khalfa and Robert Young, trans. Steven Corcoran (London: Bloomsbury, 2018), pp. 167–530.

Chapter 1 Psychoanalysis and Algerian Paradoxes

1 Sarah Haider, *Virgules en trombe* (Algiers: APIC Éditions, 2013), p. 44.
2 Amin Maalouf, *Les Désorientés* (Paris: Grasset, 2012), p. 69.
3 In November 2001, a brutal flood in the Bab El-Oued neighborhood killed 754 people and left 122 more missing (see "Huit mois après, qu'est Bab El-Oued devenu?," *L'Expression*, August 5, 2002, <https://algeria-watch.org/?p=58096>). In May 2003, a violent earthquake struck Bourmerdès, killing an estimated 2,217 people and injuring 9,085 more, not to mention countless missing people, whose numbers are always hard to determine in Algeria (see "Le dernier bilan s'élève à 2217 morts," *Jeune Indépendant*, May 27, 2003, <https://algeria-watch.org/?p=5534>).
4 Sigmund Freud, *Civilization and Its Discontents* (1929), trans. James Strachey (New York: W. W. Norton & Company, 2010), p. 17.
5 Mohammed Dib, *Dieu en barbarie* (Paris: Seuil, 1970), p. 23.
6 Dib, *L'Arbre à dire* (Paris: Albin Michel; Algiers/Hibr, 1998), p. 22.
7 Dib, *Dieu en barbarie*, p. 83.
8 *Translator's Note*: When possible, I have decided to leave this term untranslated. "*Détournement*" (literally, "turning away") is often translated in English as "diversion" or "misuse." "*Détournement de pouvoir*" is an "abuse of power" and "*détournement de fonds*" is the term used for "embezzlement," just as "*détournement d'avion*" is how one would talk about a plane being hijacked. In all these cases, one can clearly discern the detour at stake. In literary and artistic discourse, one might think about this as a

form of subversive appropriation or deliberate distortion meant to undermine a particular set of power relations. This is often how it is used throughout this work.

9 Nabile Farès, *L'Exil et le désarroi* (Paris: Maspero, 1976), p. 16.
10 Samir Toumi, *L'Effacement* (Algiers: Barzakh, 2016), p. 212. My italics.
11 Ibid., p. 213.
12 See Jacques Hassoun, *Les Contrebandiers de la mémoire* (Paris: La Découverte, 2002).
13 Dib, *Dieu en barbarie*, p. 195.
14 Amin Zaoui, *Le Dernier Juif de Tamentit* (Algiers: Barzakh, 2012).
15 Sigmund Freud, *The Future of an Illusion* (1927), trans. James Strachey (New York: Grove Press, 1989), p. 6.
16 Ibid.
17 Nabile Farès, *Le Champ des oliviers* (Paris: Seuil, 1972), p. 159. My italics.
18 Sigmund Freud, *Reflections on War and Death* (1915), trans. A. A. Brill and Alfred B. Kuttner (Project Gutenberg <http://www.gutenberg.org/ebooks/35875>).
19 Dib, *Dieu en barbarie*, p. 190.
20 Frantz Fanon, *The Wretched of the Earth* (1961), trans. Richard Philcox (New York: Grove Press, 2004), p. 116.
21 Mohamed Mebtoul, *La Citoyenneté en question (Algérie)* (Oran: Dar El Adib, 2013), p. 9.
22 Ibid., p. 155.
23 Dib, *Dieu en barbarie*, p. 108.
24 Mouloud Mammeri, *Le Sommeil du juste* (Paris: Plon, 1955).

Chapter 2 Colonial Rupture

1 Jacques Hassoun, "Le circuit de la haine dans la société colonial," in Anny Combrichon (ed.), *Psychanalyse et decolonisation. Hommage à Octave Mannoni* (Paris: L'Harmattan, 1999), p. 124.
2 Kateb Yacine, *Nedjma*, trans. Richard Howard (Charlottesville: University of Virginia Press, 1991), pp. 265–6.
3 Ibid., p. 233.
4 Abderrahmane Bouchène, Jean-Pierre Payroulou, Ouanassa Siari Tengour, and Sylvie Thénault (eds.), *Histoire de l'Algérie à la période colonial (1830–1962)* (Paris: La Découverte; Algiers: Barzakh, 2012), p. 24.
5 Yacine, *Nedjma*, p. 135.
6 See Gilles Manceron, *Marianne et les colonies. Une introduction à l'histoire colonial de la France* (Paris: La Découverte, 2003), p. 79.

7 The demographer Kamel Kateb estimates that, out of a population of 3 million inhabitants on the eve of the invasion, a third died between 1830 and 1875 due to massacres, crackdowns by the French army (around 825,000 deaths), epidemics, and famines. See his "Le bilan démographique de la conquête de l'Algérie (1830–1880)," in Bouchène et al. (eds.), *Histoire de l'Algérie à la période coloniale (1830–1962)*, pp. 82–8.

8 Alexis de Tocqueville, *Writings on Empire and Slavery*, ed. and trans. Jennifer Pitts (Baltimore/London: Johns Hopkins University Press, 2001), p. 135. Translation modified.

9 Cited in François Maspero, *L'Honneur de Saint-Arnaud* (Paris: Seuil, coll. "Points," 2012), p. 93.

10 Alexis de Tocqueville, *De la démocratie en Amérique* (1840), *Oeuvres*, vol. 1, chap. 10 (Paris: Gallimard, 1991), p. 427.

11 Maspero cites Saint-Arnaud in this regard: "I'm testing an idea for a major problem: let the Arabs conquer, punish, subjugate themselves. Huge advantage is that they will exterminate themselves" (Maspero, *L'Honneur de Saint-Arnaud*, p. 241).

12 Maspero also shares a note Marshal Bugeaud wrote on June 11, 1845, concerning the inhabitants of Dahra who had sought refuge in caves and were executed by Pellissier: "If these bastards retreat to their caves, let's take Cavaignac's approach: smoke them out like foxes" (quoted in ibid., p. 235). For his part, Saint-Arnaud would go so far as to suffocate entire tribes with smoke, with up to 800 people in a cave. The accounts of men, women, children, and the elderly choking on smoke and convulsing are deeply upsetting.

13 Yacine, *Nedjma*, pp. 134–6. My italics.

14 Benjamin Stora, "Quand une mémoire (de guerre) peut en cacher une autre (coloniale)," in Pascal Blanchard, Nicolas Bancel, and Sandrine Lemaire (eds.), *Fractures colonials. La société française au prisme de l'héritage colonial* (Paris: La Découverte, 2005), p. 64.

15 On this point, see the 1885 parliamentary debates between Georges Clemenceau and Jules Ferry (reprinted in *1885, le tournant colonial de la République. Jules Ferry contre Georges Clemenceau et autres affrontements parlementaires sur la conquête colonial*, introduction by Gilles Manceron [Paris: La Découverte, 2006]).

16 Cited by Michel Wieviorka, in Blanchard et al. (eds.), *Fractures coloniales*, p. 118.

17 Tocqueville, *Writings on Empire and Slavery*, p. 141.

18 "We preserved nothing of the former government of the country but the employment of the yagatan and the baton as police equipment. Everything else became French" (ibid., p. 16).

19 *1885, le tournant colonial de la République*, p. 11.

20 See Bouchène et al. (eds.), *Histoire de l'Algérie à la période coloniale (1830–1962)*, p. 64.
21 Olivier Le Cour Grandmaison, *Coloniser, exterminer. Sur la guerre et l'État colonial* (Paris: Fayard, 2005), p. 158.
22 Mehdi Lallaoui, *Algériens du Pacifique. Les déportés de Nouvelle-Calédonie* (Paris: Au nom de la mémoire, 1994; Algiers: Zyriab, 2001).
23 In 1960, during the War of Liberation, around a quarter of the population had been placed in internment camps (Le Cour Grandmaison, *Coloniser, exterminer*, p. 94).
24 Kateb Yacine writes: "Hunger led me quickly from one sidewalk to the other" (*Nedjma*, p. 73); and Jean El Mouhoub Amrouche: "There was hunger, a terrible hunger, widespread in the rural areas, those African hungers which France hadn't experienced for centuries. Not a lack of excess ... but a lack of the minimum needed for survival, so bad in even normal times that for a French person from France it would be seen as extreme misery" (Jean El Mouhoub Amrouche, *Un Algérien s'adresse aux Français. Ou l'histoire de l'Algérie par les textes, 1943–1960* [Paris: L'Harmattan, 1994; Algiers: Dar Khattab, 2013], pp. 284–5).
25 See Olivier Le Cour Grandmaison, *De l'indigénat. Anatomie d'un "monstre" juridique: le droit colonial en Algérie et dans l'Empire français* (Paris: La Découverte/Zones, 2010).
26 Nabile Farès, *La Découverte du Nouveau Monde*, Book II, *Mémoire de l'Absent* (Paris: Seuil, 1974), pp. 20 and 55.
27 Ibid., p. 17.
28 Yacine, *Nedjma*, p. 110.
29 According to the historian Gilbert Meynier, 136,000 Algerians fought against fascism during World War II, and 12,000 of those died in battle (see Gilbert Meynier, *Histoire intérieure du FLN, 1954–1962* [Paris: Fayard, 2002]).
30 Benjamin Stora, "Entre la France et l'Algérie, le traumatisme (post) colonial des années 2000," in Ahmed Boubeker, Françoise Vergès, Florence Bernault, Achille Mbembe, Nicolas Bancel, and Pascal Blanchard (eds.), *Ruptures postcoloniales. Les nouveaux visages de la société française* (Paris: La Découverte, 2010).
31 Amrouche, *Un Algérien s'adresse aux Français*, p. 58.
32 Jean El Mouhoud Amrouche, "L'éternel Jugurtha," *L'Arche*, no. 13, 1946 (Bougie: Tafat Éditions, 2016), p. 17. My italics. Jugurtha was one of the last kings of Numidia, an ancient Berber kingdom that corresponds today to Northern Algeria (see chapter 8).
33 "Not a relative the way the French understand it" (Yacine, p. 164).
34 Ibid., p. 171.
35 Farès, *Mémoire de l'Absent*, p. 179.

36 Yacine, *Nedjma*, p. 236.
37 Ibid., p. 240.
38 Ibid., p. 331.
39 Ibid., p. 253.
40 Ibid., p. 128.
41 Ibid., p. 110.
42 Maspero, *L'Honneur de Saint-Arnaud*, p. 232.
43 Yacine, *Nedjma*, pp. 171 and 249–50.
44 Sigmund Freud, *Totem and Taboo* (1919), trans. A. A. Brill (Project Gutenberg <http://www.gutenberg.org/ebooks/41214>).
45 Mostefa Lacheraf, *Des noms et des lieux. Mémoire d'une Algérie oubliée* (Algiers: Casbah, 1998), pp. 147–73.
46 Yacine, *Nedjma*, p. 169.
47 Farid Benramdane, "Algérianité et onomastique. Penser le changement: une question de nom propre," in *Insaniyat*, no. 57–8, July–December 2012 (Oran: GRASC), pp. 143–59.
48 Tocqueville, *Writings on Empire and Slavery*, p. 86.
49 Yacine, *Nedjma*, p. 252.
50 "That old pirate Si Moktar, the fake father ..." (ibid., p. 128).
51 Farès, *Mémoire de l'Absent*, pp. 28 and 198.
52 Yacine, *Nedjma*, pp. 127–8. My italics.
53 Dib, *L'Arbre à dire*.
54 Daho Djerbal, "De la dépossession du nom à l'expropriation de la terre par la carte," in *Made in Algeria. Généalogie d'un territoire*, exhibition catalogue of the MUCEM (Marseille, 2016), pp. 185–6.
55 See Nabile Farès, *Maghreb, étrangeté et amazighité* (Algiers: Koukou, 2016).
56 Mouloud Feraoun, *Le Fils du pauvre* (Le Puy: Cahiers du nouvel humanisme, 1950).
57 Mouloud Feraoun, *Le Journal (1955–1962)* (Paris: Seuil, 1962), p. 162.
58 On March 15, 1962, members from this French dissident paramilitary organization assassinated Marcel Basset, Robert Eymard, Mouloud Feraoun, Ali Hammouten, Max Marchand, and Salah Ould Aoudia. These National Education inspectors who were at a work meeting oversaw social centers that promoted literacy.
59 "Only one in five boys were in school, and one in sixteen girls. At the university of Algiers, four fifths of the students are of European origin" (Le Cour Grandmaison, *De l'indigénat*, p. 171).
60 Albert Memmi, *The Colonizer and the Colonized* (1974), trans. Howard Greenfeld (London: Earthscan Publications, 2003).
61 See Fadhma Aït Mansour Amrouche, *Histoire de ma vie* (Paris: Maspero, 1968).
62 Jean El Mouhoub Amrouche, *Chants berbères de Kabylie* (1939)

(Paris: L'Harmattan, 1986; Bougie: Tafat Éditions, 2016), pp. 50–1. My italics.

63 "To write in French is, on a much higher level, like snatching the gun out of a paratrooper's hands! It's of equal value" (ibid., p. 56).

64 Kateb Yacine, "Entretien avec Lia Lacombe," in *Les Lettres françaises, 7–13 février 1963*, reprinted in Kateb Yacine, *Le Poète comme un boxeur. Entretiens 1958–1989* (Paris: Seuil, 1994), p. 59.

65 Ibid., p. 15.

Chapter 3 Colonialism Consumed by War

1 Ferhat Abbas, *Autopsie d'une guerre. L'aurore* (Paris: Garnier, 1980), p. 273.

2 Malek Haddad, *La Dernière Impression* (Constantine: Média Plus; Algiers: Bouchène, 1989), p. 147.

3 Yamina Mechakra, *La Grotte éclatée* (Algiers: ENAG, 2000), p. 16.

4 Yacine, *Nedjma*, p. 26. Translation modified.

5 This loyalty toward an Islamic and "Arab world" is the consequence of many political factors. The FLN, for example, was financed by many Arab countries, most notably Egypt, who played a major role in the geopolitical struggle between different blocs.

6 Farès, *Le Champ des oliviers*, p. 126.

7 Lucien de Montagnac, *Lettres d'un soldat. Neuf années de campagnes en Afrique* (Paris: Plon, 1885; re-edited by Christian Destremeau, 1998), p. 153 (quoted in Manceron, *Marianne et les colonies*, p. 164).

8 Albert Camus, *The First Man*, trans. David Hapgood (New York: Alfred A. Knopf, 1995), p. 141.

9 Mechakra, *La Grotte éclatée*, p. 22.

10 Ibid., p. 38.

11 Ibid., p. 37.

12 Ibid., p. 31.

13 Ibid., p. 151.

14 Ibid., p. 43.

15 Ibid., p. 43.

16 Ibid., p. 79.

17 Farès, *Mémoire de l'Absent.*

18 Mechakra, *La Grotte éclatée*, p. 124.

19 Louisette Ighilahriz, *Algérienne. Récit recueilli par Anne Nivat* (Paris: Fayard/Calmann-Lévy, 2001).

20 Frantz Fanon, *A Dying Colonialism* (1965), trans. Haakon Chevalier (New York: Grove Press, 1994), p. 65.

21 Kateb Yacine, "Le cadavre encerclé," in *Le Cercle de représailles* (Paris: Seuil, 1959).
22 Daniel Mesguich, "La guerre d'Algérie en moi n'est qu'un grand trou," in *Une enfance dans la guerre, Algérie 1954–1962. Textes inédits recueillis par Leïla Sebbar* (Saint-Pourçain-sur-Sioule: Bleu autour, 2016), p. 197.
23 Benjamin Stora, *La Gangrène et l'oubli. La mémoire de la guerre d'Algérie* (Paris: La Découverte, 1991), p. 8.
24 *Translator's Note*: *Harkis* were Algerian Muslims who were recruited to fight alongside the French army during the War of Liberation. By some estimates, 150,000–200,000 Algerians fought on behalf of the French. In Algeria, the term became synonymous with "traitor" and was widely applied to anyone suspected of harboring pro-French sentiments.
25 Kateb Yacine, "Un jardin parmi les fleurs," *Esprit*, July–December 1962, p. 774.
26 Yacine, *Nedjma*, p. 233.
27 Yacine, *Le Poète comme un boxeur*, p. 56.
28 Chawki Amari, *Le Faiseur de trous* (Algiers: Barzakh, 2007). The psychoanalyst Serge Leclaire (1924–94) has also analyzed this very telling story that speaks to the place and function of the hole in the psychic economy.
29 Farès, *Mémoire de l'Absent*, p. 98. My italics.
30 These words reappear three times in the novel, as Farès asks: "How can we come to grips with why a country kills, when what I know is its gentle voyage under the kelp, the smell of salt on seaside promenades, the taste of the sky and its winking mirrors, right there, dispersed across the waves in a multitude of shattered suns" (ibid., p. 108).
31 Ibid., p. 29.
32 Ibid., p. 27.

Chapter 4 Colonialism's Devastating Effects on Post-Independence Algeria

1 Jean El Mouhoub Amrouche, *D'une amitié. Correspondance Jean Amrouche-Jules Roy (1937–1962)* (Aix-en-Provence: Édisud, 1985), p. 104.
2 Abbas, *Autopsie d'une guerre*, p. 329.
3 Amrouche, "L'éternel Jugurtha," p. 17. My italics.
4 Ibid. My italics.
5 See Baldine Saint-Girons, "L'homme intérieur et le désir du dedans,"

in Jacky Pigeaud (ed.), *L'Intérieur. XXes entretiens de La Garenne Lemot* (Rennes: PUR, 2017).

6 Sigmund Freud, *Group Psychology and the Analysis of the Ego* (1922), trans. James Strachey (Project Gutenberg <http://www.gutenberg.org/ebooks/35877>).

7 Amrouche, "L'éternel Jugurtha," pp. 22–3.

8 Freud, *Group Psychology and the Analysis of the Ego*. Translation modified.

9 Amrouche, "L'éternel Jugurtha," p. 38. My italics.

10 *Translator's Note*: In English in the original.

11 Ibid., p. 39.

12 Ibid., p. 107.

13 Memmi, *The Colonizer and the Colonized*, p. 195.

14 Amrouche, *Un Algérien s'adresse aux Français*, p. 115.

15 Ibid., p. 117.

16 See Alice Kaplan, *Looking for The Stranger: Albert Camus and the Life of a Literary Classic* (Chicago: University of Chicago Press, 2016). *Translator's Note*: The first syllable of Mersault's name evokes both the sea ("*mer*") and/or the figure of the mother ("*mère*") whereas in the second syllable one can hear the word for "idiot" ("*sot*") and/or "jump" ("*saut*"). The alternate spelling evokes murder ("you murder, he murders" for "*meurs/meurt*").

17 It remains to be seen who dies when "I" and "He" are murdered under colonialism. This point cannot be overstated: is it a question of killing the Other within one's self (the "He" of "I"), in this colonizer/colonized dynamic? Or is it the other way around, with the subject of the Other being killed (the "I" of "He")? Clearly, in both cases, what is killed is any form of alterity for colonialism's two protagonists.

18 Amrouche, *Un Algérien s'adresse aux Français*, p. 407.

19 Ibid., p. 433. My italics.

20 Ibid., p. 434. My italics.

21 Claude Rabant, *Inventer le réel, le déni entre perversion et psychose* (Paris: Hermann, 2011), p. 91.

22 Albert Memmi, *Decolonization and the Decolonized*, trans. Robert Bononno (Minneapolis: University of Minnesota Press, 2006), pp. 55–6. My italics.

23 Paul-Laurent Assoun, *Le Préjudice et l'Idéal. Pour une clinique sociale du trauma* (Paris: Anthropos, 1999), p. 16.

24 Amrouche, *Un Algérien s'adresse aux Français*, p. 115.

25 Assoun, *Le Préjudice et l'Idéal*, p. 108. My italics.

26 Kamel Daoud, *The Meursault Investigation*, trans. John Cullen (London: Oneworld Publications, 2015).

27 Jacques Lacan uses this word to designate the complexity of the

relations between the interior private sphere and the exterior private sphere. It suggests that the private sphere [*l'intime*] is an external manifestation, but also that the exterior shapes the interior. The private sphere would thus designate the complexity of this dynamic relation between the inside and outside in the formation of subjectivities.

Chapter 5 Fratricide: The Dark Side of the Political Order

1 Sigmund Freud, *Psychopathology of Everyday life* (1901), trans. Anthea Bell (London: Penguin, 2002), p. 203.
2 Freud, *Reflections on War and Death*.
3 Mohammed Harbi, *Une vie debout. Mémoires politiques*, Vol. 1, *1943–1962* (Paris: La Découverte; Algiers: Casbah, 2001), p. 224.
4 "The original lost paradise, which is increasingly lost to us," writes Benjamin Stora, "is coming alive again with religion. Islam fuels and props up the populist side of the nationalist ideology" (Stora, *La Gangrène et l'oublie*, p. 130).
5 Harbi, *Le FLN, mirage et réalité*, p. 305.
6 Ferhat Abbas, *L'Indépendence confisquée* (Paris: Flammarion, 1984).
7 See Benjamin Stora, *Messali Hadj* (Paris: Fayard, collection "Pluriel," 2012).
8 Meynier, *Histoire intérieure du FLN, 1954–1962*, p. 264.
9 "This wasn't a matter of a wholesale elimination of civilian 'traitors' as some have argued, but rather a policy of executing high-ranking officers, Algerian *djounoud* and members of *nizam* under the direction of *wilayas* who believed they were plotting against the ALN and the *tawra*, which needed to be eliminated. ... This purge was responsible for several thousand deaths between 1958 and 1962" (ibid., pp. 430–1). [*Translator's Note*: *Djound* (*djoundi* in the singular) refers to Algerian Muslim soldiers who fought for Independence during the Algerian War. *Wilayas*, or regions, refers in this case to the territories designated by the ALN during the war.]
10 Gilbert Meynier has focused on this: "The FLN chased out deviants and traitors of all stripes. The worst punishments were reserved for them, and they were often executed. ... From 1955 to 1962, the list is long (in the thousands) of people assassinated for treason or failure to comply or even simply by what was perceived as a slight to the FLN, and this isn't including internal political assassinations, nor the victims of the purges carried out

by the leaders of the ALN within its own organization, which likely added another several thousand victims" (ibid., p. 218). He adds that "the execution orders were usually issued from high up without debate. Even to ask questions about what was being ordered, including simple demands for clarification, *was generally felt by the leaders to be a sign of non-compliance. This was considered reprehensible behavior that invited suspicion*" (p. 261, my italics).

11 Jean El Mouhoub Amrouche, *Journal (1928–1962)*, ed. and introducted by Tassadit Yacine Titouh (Paris: Non Lieu, 2009), pp. 163 and 300.

12 The massacres committed by the OAS instilled terror in all the civilian populations (Muslim, Jewish, and European). Some of the atrocities committed as soon as the Évian Accords were signed remain to this day shrouded in mystery, as each camp, OAS and FLN, blames the other for these crimes. "Who's killing whom?" is already inscribed in history.

13 Meynier, *Histoire intérieure du FLN, 1954–1962*, p. 264.

14 Harbi, *Le FLN, mirage et réalité*, p. 157.

15 Ibid.

16 Ibid., p. 158. Mohammed Harbi further develops this point, stating that "the day when we'll be brave enough to study without bias the civil strife between Algerians, we'll conclude without any doubt that this tragedy was not inevitable and that the responsibility of the FLN, in the turn of events, is not negligible. ... Many have characterized the opposition between the FLN and the MNA as a clash of politics. In fact, it is better seen as a rivalry between two movements seeking absolute power, both after the same thing but with different approaches" (p. 160).

17 For more on this matter, see the book by the historian Mahfoud Kaddache, *Et l'Algérie se libéra, 1954–1962* (Paris: Paris-Méditerranée; Algiers: EDIF 2000, 2003).

18 Ibid., p. 72.

19 Ibid., p. 129.

20 Abbas, *Autopsie d'une guerre*.

21 Abbas, *L'Indépendence confisquée*, p. 37.

22 See Tarik Khider, *L'Affaire Khider. Histoire d'un crime d'État impuni* (Algiers: Koukou, 2017). Khider tells the story of his father's political assassination by "his revolutionary brothers-in-arms": "For the Algerian regime, physical removal became the answer for settling political conflicts. Many figures, who were either declared enemies or didn't have the fortune of belonging to the dominant clan, were taken out; others met their end in suspect circumstances, such as Colonel Saïd Adid ('suicide'), Colonel Abbés (killed in a car

accident), Colonel Chaâbani ('executed' after a sham trial), Colonel Amirouche (taken down in a random raid), Mohamed Khmisti (assassinated by an 'insane' man), Mohamed-Seddik Benyahia (died after a plane crash), Ahmed Medeghri (three bullets in the head, ruled a suicide), Mohamed Boudiaf (assassinated on live TV by a member of his own security squad), Ali Mécili (assassinated by a thug hired by the *Sécurité Militaire*), Matoub Lounés (riddled with bullets by a mysterious fighter)" (p. 9).

23 Houari Boumediene is the name of the patron saint of the city of Oran which he adopted as his *nom de guerre*. His true birth name is Mohamed Boukharouba, and he was born in 1932. It is interesting to note the confluence of war and religion in this name used to confront the occupying power.

24 As Mohammed Harbi recounts, "Overlapping personal interests start to trump political affinities. No one has a coherent strategy for the present or the future. *The problem is to survive. Everyone is suspicious of everyone and is concerned with potentially acting to neutralize the threat of the other*" (Harbi, *Le FLN, mirage et réalité*, p. 204; my italics).

25 This detail is related by Ali Zamoum in *Le Pays des hommes libres. Tamurt Imazighen. Mémoires d'un combattant algérien (1940–1962)* (Grenoble: La Pensée sauvage, 1998), p. 292.

26 Abbas, *L'Indépendence confisquée*, p. 57.

27 See Khider, *L'Affaire Khider*; and Belaïd Abane, *Nuages sur la Révolution. Abane au coeur de la tempête* (Algiers: Koukou, 2015).

28 See, among others, Rachid Benyellès, *Dans les arcanes du pouvoir, 1962–1999* (Algiers: Barzakh, 2017); and Moktar Mokhtefi, *J'étais français-musulman* (Algiers: Barzakh, 2016).

29 Harbi, *Le FLN, mirage et réalité*, p. 204. My italics.

30 Freud, *Totem and Taboo*.

31 Ibid., n. 21. My italics.

32 Gilles Carpentier, "Préface," in Kateb Yacine, *Nedjma* (Paris: Point, 1996), p. 10.

33 Yacine, *Nedjma* [English translation], p. 240.

34 Ibid., p. 231.

35 "Come and take a look at the shape-shifting vacuum of being as I've called out man's most complete subjugations in order to free his son for the land of men" (Farès, *Mémoire de l'Absent*, p. 95).

36 Kateb Yacine, *Le Polygone étoilé* (Paris: Seuil, 1966), p. 9. My italics.

37 *Translator's Note*: In English in the original.

38 Yacine, *Nedjma*, p. 161.

Chapter 6 The Internal War of the 1990s

1 Freud, *Civilization and Its Discontents*, p. 61.
2 Waciny Laredj, *Les Ailes de la reine*, trans. Marcel Bois (Paris: Sindbad, 1993), pp. 40 and 74.
3 That the revelation of the Qur'anic message took place in Arabic is one thing, but it's another thing to abolish altogether the history of this language before the emergence of Islam.
4 Rachid Mimouni, *Une peine à vivre* (Paris: Stock, 1991).
5 Ibid., p. 11.
6 For more on this issue, see "De quelques ravages de la langue Une," in Karima Lazali, *La Parole oubliée* (Toulouse: Érès, 2015).
7 Mimouni, *Une peine à vivre*, p. 246.
8 Mansour Kedidir, *La Nuit la plus longue* (Algiers: APIC, 2015), p. 74. Kedidir had long served as an examining magistrate, attorney general, and chief of staff to the head of government. In this novel, he provides meticulous descriptions of the Internal War of the 1990s and the ravage it caused, from the atrocities committed by armed Islamists to the means used to spread terror. One of his characters, a teacher who has gone to Afghanistan, at one point says, "We know that the Arab regimes took advantage of this opportunity to get rid of our brothers. *But the day will come when we'll return as victors and we'll make the earth rattle under their feet*" (ibid., my italics).
9 Hassane Zerrouky, *La Nébuleuse islamiste en France et en Algérie* (Paris: Éditions 1, 2002).
10 This party's political platform argued to ban co-education and public beaches and to require that the veil be worn. Of course, among many other things, it also sought to apply Sharia law and adopt an aggressive policing of behavior.
11 This was the term used by Ali Benhadj, the FIS leader, in 1990.
12 Arezki Mellal, *Maintenant, ils peuvent venir* (Algiers: Barzakh; Arles/Actes Sud, 2002), p. 98.
13 "Twenty-four years after the cowardly assassination of Si Tayeb El Watani, we, his children, continue to firmly reject the claim of an isolated act put forward by his assassins, Belkhier, Nezzar, Toufik, and Smaïn, the same people behind his return to Algeria. These assassins thought they were dealing with a senile man who could become their puppet, but they were wrong. The man of November, to their great surprise, demonstrated a deep understanding of his own people, which allowed him to grasp the reality of the country and its youth" ("Assassinat de Boudiaf: Son fils demande la révision du procès," <http://dia-algerie.com/assassinat-de-boudiaf-fils-demande-revision-proces/>, June 2016).

14 Among others, the example in France of the Kouachi brothers, who carried out the *Charlie-Hebdo* attack in January 2015, attests to the engagements of brotherhoods in the armed Islamist movement. The notion of "terrorist," widely applied in Algeria and France to designate these Islamists, only works to depoliticize these decidedly political movements with an international reach.
15 Rachid Mimouni, *La Malédiction* (Paris: Stock, 1993).
16 Ibid., p. 266.
17 Ibid., pp. 279–80.
18 Kedidir, *La Nuit la plus longue*, p. 357.
19 "Mohamed Saïd, Hassan Hattab, and other Islamic activists were widely believed to be sons of *harkis*. Whether this is true or not matters little. This simply proves that the belief itself had a real hold over people and had an impact on contemporary society" (Abderrahmane Moussaoui, *De la violence en Algérie. Les lois du chaos* [Arles: Actes Sud; Algiers: Barzakh, 2006], p. 128).
20 Ibid., p. 256.
21 "You must understand that in this case, the extremist is like a rabid animal. He is no longer in control of himself; rape, pillaging, and murder are acts that would only bring him closer to God as long as he believes that he performs them in His name. We're dealing with a case of atypical rage." This is followed by a long dialogue between two of the novel's characters who are looking for an appropriate cure (Kedidir, *La Nuit la plus longue*, p. 177).
22 Moussaoui, *De la violence en Algérie*, p. 256.
23 Fethi Benslama, "L'idéal blessé et le surmusulman," in *L'Idéal et la cruauté. Subjectivité et politique de la radicalisation* (Paris: Lignes, 2015), p. 11.
24 Amin Maalouf, *In the Name of Identity: Violence and the Need to Belong*, trans. Barbara Bray (New York: Arcade, 2012), p. 66.
25 Quoted in Zerrouky, *La Nébuleuse islamiste en France et en Algérie*, p. 143. My italics.
26 Ibid., p. 271.
27 Nicole Loraux, *The Divided City: On Memory and Forgetting in Ancient Athens* (New York: Zone Books, 2006), p. 25.
28 Nabile Farès, *Il était une fois, l'Algérie. Conte roman fantastique* (Achab: Tizi Ouzou, 2010), p. 109.

Chapter 7 State of Terror and State Terror

1 Kedidir, *La Nuit la plus longue*, p. 135.
2 Farès, *Il était une fois, l'Algérie*, p. 21.
3 Zerrouky, *La Nébuleuse islamiste en France et en Algérie*, p. 332.

4 Ibid., p. 110.
5 Paul-Laurent Assoun, *Tuer le mort. Le désir révolutionnaire* (Paris: PUF, 2016).
6 I refer the reader here to the scrupulous and astonishing description of the state of terror provided to a German colleague at the end of 1993 by the anti-Islamist journalist Saïd Mekbel, one year before his assassination by the army, if we believe his own prediction, or by the Islamists, according to the state security apparatus (Monkia Borgmann, *Saïd Mekbel, une mort à la lettre* [Paris: Téraèdre; Beirut: Dar al-Jadeed, 2008; reprinted in Algeria: Tizi-Ouzou: Éditions Frantz Fanon, 2016]).
7 Tahar Djaout, *The Bone Seekers*, trans. Marjolijn de Jager (New Orleans: Diálogos Books, 2018).
8 Mellal, *Maintenant, ils peuvent venir*, p. 91.
9 Ibid., p. 100.
10 Ibid., p. 108.
11 Ibid., p. 199.
12 Ibid., pp. 200–1.
13 Ibid.
14 Ibid.
15 Ibid., p. 202.
16 In *The Ego and the Id* (1923), Freud considers whether the unconscious can exist without repression.
17 Farès, *Il était une fois, l'Algérie*, p. 52.
18 Sándor Ferenczi, *Le Traumatisme* (1934) (Paris: Payot, coll. "Petite Bibliothèque Payot," 2006), p. 57.
19 Ibid., p. 63. My italics.
20 Ibid., p. 67.
21 See Giovanna Borrador, *Philosophy in a Time of Terror: Dialogues with Jürgen Habermas and Jacques Derrida* (Chicago: University of Chicago Press, 2003).
22 Ferenczi, *Le Traumatisme*, p. 67.
23 Ibid., p. 145. My italics.
24 This figure has been forcefully contested by France, and remains a matter of debate between historians.
25 Freud, *Reflections on War and Death*.
26 See the report of the Collective of Families of the Disappeared in Algeria, *Les Disparitions forcées en Algérie, un crime contre l'humanité, 1990–2000*, <http://www.algerie-disparus.org>, February 2016.
27 There are very few fictional treatments of this matter, apart from the very beautiful book by Wahiba Khiari, *Nos silences* (Tunis: Elyzad, 2009).
28 Gilou García Reinoso, "Tuer la mort," in Heitor O'Dwyer de

Macedo (ed.), *Le Psychanalyste sous la terreur* (Paris: Matrice, 1988), p. 171.

Chapter 8 Legitimacy, Fratricide, and Power

1 Mimouni, *La Malédiction*, p. 268.
2 Sallust, *The War with Jugurtha*, trans. John C. Rolfe (Cambridge, MA: Loeb Classical Library, 1931), p. 206.
3 Ibid., p. 205. My italics.
4 Ibid., p. 149. Translation modified.
5 Ibid., pp. 151–2. Translation modified.
6 Ibid., pp. 160–1.
7 Houaria Kadra, *Un Berbère contre Rome* (Paris: Arléa, 2005).
8 Sallust, *The War with Jugurtha*, pp. 204, 206.
9 The issue of unpunished crimes, as shown, is a thorny matter that continues to plague the Algerian Republic. The dismissal of crimes benefits both the Islamists and state agents. And this situation only further spreads confusion and makes it impossible to clearly establish responsibility and to chart out a history or form of governance. It is interesting to note that two warring groups find themselves put on the same level by the "law of national reconciliation" to the detriment of the general public, which is eagerly seeking to understand the situation and find some form of solace. This law does little to quell the state of terror, permitting, instead, corruption and perversion to take place undercover. In this context, it is hard for subjects to see the law as an independent force ensuring their safety and well-being since nothing seems to stand in the way of people's anarchic impulses and their deeply criminal and incestuous nature.
10 Ibid, p. 379. My italics.
11 Jacques Lacan, *The Other Side of Psychoanalysis: The Seminar of Jacques Lacan, Book XVII*, ed. Jacques-Alain Miller, trans. Russell Grigg (New York: W. W. Norton, 2007), p. 114.
12 Jacques Lacan, *Transference: The Seminar of Jacques Lacan, Book VIII*, ed, Jacques-Alain Miller, trans. Bruce Fink (Cambridge: Polity, 2015), p. 289: "And in truth, we have here a theme that would be well worth considering in the genesis of what we call colonialism, which is the theme of émigrés who did not simply invade colonized countries but who broke new ground. This resource given to all the *lost children of Christian culture* would certainly be worth isolating as an ethical mainspring that we would be wrong to neglect at the present moment where we are weighing its consequences" (my italics).

13 Cited in Le Cour Grandmaison, *Coloniser, exterminer*, p. 13. My italics.
14 Todd Shepard, “Plus grande que l’Hexagone”, in *Made in Algeria. Généalogie d’un territoire*, p. 166.
15 Albert Camus, “A Short Guide to Towns without a Past”, in *Lyrical and Critical Essays*, trans. Ellen Conroy Kennedy (New York: Knopf, 1968), p. 143.
16 Louis Bertrand, *Africa* (Paris: Albin Michel, 1933), p. xviii (quoted in Farès, *Maghreb, étrangeté et amazighité*, p. 104).
17 Le Cour Grandmaison, *Coloniser, exterminer*, p. 112.
18 For more on this matter, see Catherine Brun and Todd Shepard (eds.), *Guerre d’Algérie, le sexe outragé* (Paris: CNRS Éditions, 2016).
19 Bertrand, *Africa*, p. xviii. My italics.
20 Ibid.
21 Albert Camus “The Wind at Djemila,” in *Lyrical and Critical Essays*, pp. 73–4. My italics.
22 Repetition and re-enactment function on different levels in the psyche. The first, stemming from repression, allows what is repressed to return in different forms; it contains the potential for creativity between the repressed original text and the one that returns; from one to the next, rewriting takes place. Re-enactment, however, works on another level: what returns outside of memory comes back the same; no rewriting is possible. Re-enactment adds nothing new to the erased text nor opens up the possibility for a future rewriting. In the context of repression, the psychoanalyst becomes an archaeologist (Freud) who seeks to gather and reassemble the scattered pieces of what took place. In the case of *forclusion*, or foreclosure, the return occasioned by re-enactment is utterly detached from the context in which it was formed: history, affects, thoughts, relations, and so on. With this creation of blank space, the psychoanalyst is a participating witness who lends her psyche to a writing that never happened and which remains precluded for the thinking subject.
23 Héla Ouardi, *Les Derniers Jours de Muhammad* (Paris: Albin Michel, 2017).
24 Ibid., p. 18.
25 Ibid., p. 58.
26 For an analysis of this myth, see Karima Lazali, “L’enfant endormi,” in *La Parole oubliée*.
27 Ouardi, *Les Derniers Jours de Muhammad*, p. 99.
28 Ibid., p. 139.
29 Ouardi adds: “The recurring theme of discord shows to what extent the obsession of an internal violence which must constantly

be controlled represents a historical blind spot for Muslims" (ibid., p. 182).

30 In this respect, it is worth noting that dramas around inheritance regularly escalate into full-blown crises in Algeria. These often cause serious rifts in families, who find themselves caught in a system of indivision (*fridha*) that strives toward unity but usually splits up into warring factions. In other cases, the divisions are mutually drawn, but at the cost of irremediably fracturing family ties. The women (and girls) at times prefer to give up their share of the inheritance in order to avoid, as they say, the obscenity of these fights. In reality, insufficiently protected by secular laws, they are called upon by their religion to restrain from getting involved due to their nominal share of the inheritance. (Under Sharia law, the daughter's inheritance is half of the son's.) This religious law of inheritance is still enforced today in Algeria. Is it a question of men and the political order looking after the brothers' pleasure by excluding women?

Chapter 9 Getting Out of the Colonial Pact

1 Haddad, *La Dernière Impression*, p. 181.
2 Farès, *L'Exil et le désarroi*, p. 96.
3 Salim Bachi, *Le Chien d'Ulysse* (Paris: Gallimard; Algiers: Barzahk, 2001), p. 287.
4 *Translator's Note*: In English in the original.
5 Fanon, *A Dying Colonialism*, p. 26.
6 Fanon, *The Wretched of the Earth*, p. 139.
7 See Fanon, "Psychiatric Writings."
8 Fanon, *The Wretched of the Earth*, p. 5.
9 Farès, *Le Champ des oliviers*, p. 126.
10 Ibid., p. 37.
11 Sándor Ferenczi, "Two Types of War Neurosis" (1916/17) in *Further Contributions to the Theory and Technique of Psycho-Analysis* (n.p.: Read Books, 2013), p. 192.
12 Fanon, *A Dying Colonialism*, pp. 52–3.
13 This is not a metaphor for the situation of the contemporary President of the Republic, but a haunting reality.
14 Victor Hugo, "Discours sur l'Afrique," in *Actes et paroles* (Paris: Robert Laffont, 1992; quoted by Manceron, *Marianne et les colonies*, p. 183).
15 Inès Amroude, "L'effrayant cauchemar continue," *Le Midi*, September 20, 2015, <frama.link/wELUFpmr>.
16 Ghania Lassal, "Phénomène du kidnapping d'enfants: les parents gagnés par l'angoisse," *El Watan*, October 28, 2015.

17 Nadia Bellil, “Enlèvement d’enfants, des statistiques et un phénomène à cerner,” *Reporters*, August 7, 2016.
18 Djaout, *The Bone Seekers*, p. 72.
19 Ibid., p. 10.
20 A newspaper article claims that the first cases of missing children occurred in 2000, while another attributes the first occurrence to 2003 with the abduction in Algiers (in the Ould Fayet neighborhood) of a child named Habiba, who went missing on December 14, 2003.
21 Farès, *Mémoire de l’Absent*, p. 123.
22 Djaout, *The Bone Seekers*, pp. 158–9. My italics.

Conclusion: Ending the Colonial Curse: Lessons from Fanon

1 See Fanon, *The Wretched of the Earth*.
2 Ibid.
3 Fanon, “The Phenomenon of Agitation in the Psychiatric Milieu: General Considerations, Psychopathological Meaning,” in *Alienation and Freedom*, p. 443. He writes: “Something else is required for a hallucination to appear, critically a breakdown of the real world” (ibid.).
4 “If he is left to himself, a neurotic is obliged to replace by his own symptom formations from which he is excluded. He creates his own world of imagination for himself, his religion, his own system of delusions, and thus recapitulates the institutions of humanity” (Freud, *Group Psychology and Analysis of the Ego*).
5 Fanon, *The Wretched of the Earth*, pp. 19–20.
6 Matthieu Renault has commented on this aspect of Fanon’s thinking: “Fanon’s practice should be understood as a practice of decolonizing knowledge, which is to say, as a series of epistemic shifts, which a purely biographical-historical approach does not allow us to discover, since ‘situating Fanon in his historical context’ erases the de-contextualizing impulse in his work, an impulse that allows Fanon to push European thought beyond its borders” (Matthieu Renault, *Frantz Fanon, de l’anticolonialisme à la critique postcoloniale* [Paris: Éditions Amsterdam, 2011], p. 36).
7 Frantz Fanon, “Letter to the Resident Minister” (December 1956), in *Alienation and Freedom*, p. 434.
8 Frantz Fanon, “Day Hospitalization in Psychiatry” (1959), in ibid., p. 497.
9 Jacques Lacan, “Presentation on Psychical Causality,” in *Écrits:*

The First Complete Edition in English, trans. Bruce Fink (New York: W. W. Norton, 2006), p. 144.

10 Frantz Fanon, *Black Skin, White Masks* (1952), trans. Richard Philcox (New York: Grove Press, 2008), p. 204.

11 Ibid., p. 205.

Index